The Malaria Capers

BY THE SAME AUTHOR

New Guinea Tapeworms and Jewish Grandmothers

The Thorn in the Starfish

Who Gave Pinta to the Santa Maria?

Federal Bodysnatchers and the New Guinea Virus

The Malaria Capers

More tales of parasites and people, research and reality

Robert S. Desowitz

W · W · NORTON & COMPANY

NEW YORK · LONDON

Printed in the United States of America.

First published as a Norton paperback 1993.

The text of this book is composed in Caledonia,
with the display set in Baskerville.
Composition and manufacturing by the Maple-Vail Book Manufacturing Group.

Library of Congress Cataloging-in-Publication Data
Desowitz, Robert S.
The malaria capers : more tales of parasites and people, research and reality / by Robert S. Desowitz.
p. cm.
Includes index.
1. Malaria—History. 2. Kála-Azár—History. I. Title.
RA644.M2D53 1991
616,9′362′0096—dc20 90-28290
ISBN-13: 978-0-393-31008-5
ISBN-10: 0-393-31008-6
W.W. Norton & Company, Inc.
500 Fifth Avenue, New York, N.Y. 10110
www.wwnorton.com

W.W. Norton & Company Ltd.
Castle House, 75/76 Wells Street, London WIT3QT

8 9 0

For Carrolee

Contents

Acknowledgments

I AM INDEBTED to my wife Carrolee, and to Dr. Carol Jenkins of the Papua New Guinea Institute of Medical Research for reading the manuscript and for their valuable criticisms and suggestions. I thank Dr. Michael Alpers, Director of the Papua New Guinea Institute of Medical Research, for his friendship, advice, assistance, and many kindnesses during my visits to Papua New Guinea.

I am also indebted to the Rockefeller Foundation for a scholar-in-residence award at their Study Center, the Villa Serbelloni, in Bellagio, Italy, where I began writing this book.

Introduction

Chapter 1

Nothing Really Changes in Salata

WHEN WE RETURNED to Salata, there was now a track and the four-wheel-drive vehicle took us within an easy walking mile. It was a better day than that of twenty years before, when I had made the six-mile march with a retinue of carriers to begin a week-long study of malaria.[1] The village chief who now met us was not in the "sing-sing" dress and maquillage of his predecessor, nor did he proffer his lowered hand for the traditional scrotal greeting of the Torricelli Range people.

The years had imposed other changes. The soaring spirit houses, the *haus tambarans*, were gone, destroyed at the insistence of the ever-growing influence of fundamentalist Christian missionaries. Later, a villager would take me, conspiratorially—one heathen to another—to a *haus tambaran* deep in the jungle; hidden from the missionaries and their God. It was a sorry bush structure with little of the superb artistry and artistic value of their traditional houses of worship. One old man peered at me and said he remembered my last visit twenty years ago, but he thought I had changed "lik lik." Obviously, he must have been thinking

1. See chapter 14, "Unseemly Behavior," in Robert S. Desowitz, *New Guinea Tapeworms and Jewish Grandmothers: Tales of Parasites and People* (New York: W. W. Norton, 1981).

of someone else. No other old-timers remembered me. There weren't many old-timers. Life is not long in New Guinea's East Sepik region.

In Salata, that first day of return, we worked through the heat of the long day: Ray Spark in a deer stalker hat, his butterfly net at the ready beside him, taking blood films and chatting up the villagers in flawless pidgin; George Anian, a Papua New Guinean research associate, the painless blood drawer, taking serum samples; myself, palpating abdomens to feel for the malaria-enlarged spleens—no problem in children but more difficult in the muscular, ticklish adults.

"Breathe in" *(holim win)*—giggle, giggle.

By the end of the day we knew that nothing had changed in Salata; malaria was as it had been twenty years ago. Later examination of the stained blood films in the laboratory not only confirmed our first assessment based on the percentage of people with enlarged spleens but also revealed that malaria was now more prevalent, less treatable, less controllable than in 1962. Moreover, the adults, who in former times were clinically immune, had lost that hard-earned protection and now were frequently feverish. An adjacent control demonstration village, Bumbita, in which malaria had been reduced to negligible endemicity in 1964 by intensive spraying with DDT and mass antimalarial drug administration, was now like Salata—highly malarious. Nor was it only malaria that remained unwholesomely unchanged. There was a new generation of snotty-nosed kids with their upper respiratory tract infections. Hardly a clean skin could be seen in the village; almost everyone still had the thickened, horny "puk-puk" (crocodile) skin of the tinea imbricata fungus infection.

There are many "Salatas" throughout the tropical world

where health and health systems have deteriorated during the past twenty-five years. These are villages in the poor nations that have not benefited from the massive largesse of Western charity. These are Salatas which have not benefited from the medical research, research of unimpeachable intellectual quality, carried out in Western biotechnology establishments. Tropical Africa is the "Great Salata." It has been virtually untouched by, or responsive to, antimalarial and other public health programs. In 1984, the year we were making our resurvey of Salata, the Nigerians published their own plaintive report: "Despite health delivery care, parasite [malaria] rates in Nigeria are the same as in 1934. Only 1% of the population is covered by vector [mosquito] control." Even so, from 1934 to 1954 tropical Africa, with all its sleeping sickness, its 40 percent childhood mortality from malaria and malaria-related disease, its schistosomiasis, its tuberculosis, its meningitis of many kinds, and its dysenteries of many kinds, was relatively healthy. "Relatively" because now Africa has AIDS. The virus is killing off great masses of Africa's peoples.

Africa is the continent of sorrows where one disease feeds on another and it is where malaria is bringing AIDS to its children. Young children are particularly at risk to becoming infected with acute life-threatening malaria. That threat is now even greater because of the recent appearance of drug-resistant strains. One of the most common consequences of childhood acute malaria is an anemia so severe that a blood transfusion is required to save the patient's life. Dying children are brought to hospital for blood transfusion. In hospital they get transfusions from donors whose blood has not been tested for antibody to the AIDS virus. No such screening can be performed in the simple laboratories of African health centers despite the need—over 30 percent of the adult population may have the AIDS virus in their blood. In this way the children come into the hospital with malaria and leave with AIDS.

Thus, during the past two decades when biotechnology has made so many stunning advances, the health of tropical peoples has worsened. Eradication and control schemes have collapsed. Old, proven therapies have become impotent in the face of drug-resistant microorganisms. New, affordable, non-toxic chemotherapeutics have not been developed; a drugs-for-profit pharmaceutical industry gives low priority to the diseases of poor people. So, too, with insecticides; arthropod-borne diseases (malaria, yellow fever, sleeping sickness, filariasis, dengue fever, Chagas' disease, Japanese B encephalitis—to name a few), once controllable by long-acting insecticides (notably DDT), have lost their power by reason of insecticide-resistant insects and insecticide-resistant health authorities. Expertise has been lost; the last generation of truly experienced "field hands" are leaving the scene, lost to age and disuse. They are being replaced both in the West and in the research centers of the tropics by the "molecular types," more concerned with the exquisite intellectual challenges of modish science than with seeking practical solutions. The razzledazzle and promise of biotechnology have led Third World health officials to expect the quick fix—the malaria vaccine "just around the corner," the genetically altered mosquito that yesterday's press release proclaims will be the last word in controlling vector-borne diseases; and, confusing diagnosis with cure, the DNA probe techniques to detect parasites even at clinically insignificant levels.

That is the weft of our story. There is an imbalance, a discontinuity between research and reality. This is an imbalance that has inhibited improvement in the health of tropical peoples; but in addition, I believe it has actually contributed to the deterioration of health. The warp is formed by the story of two great tropical diseases, kala azar (visceral leishmaniasis) as it occurs in the Indian subcontinent, and malaria as it now occurs, and where it formerly occurred, throughout its near-worldwide endemic dominion. These diseases will serve as examples of the tropical world's state of health. We will follow each of these insect-transmitted infections through the course of their natural history, human history, and the historical events surrounding their elucidation by sometimes great, sometimes petty, and sometimes venal scientists.

Let us begin by imagining ourselves in an impoverished kala azar-stricken village. It is April and . . .

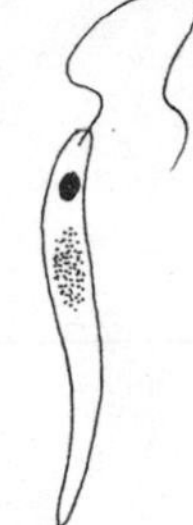

I

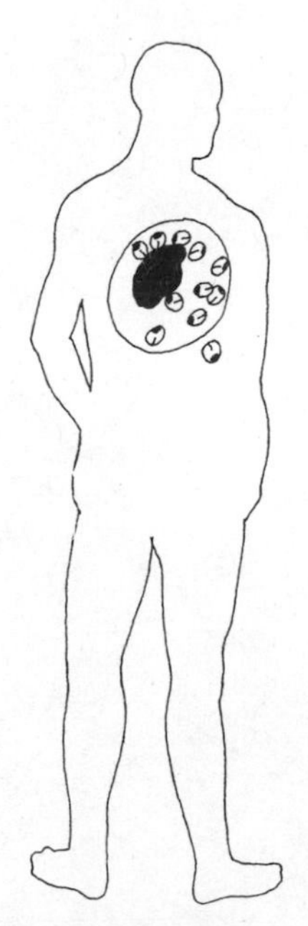

Kala Azar: The Long Anguish of the Black Sickness

Chapter 2

A Child Dies in a Small Village

THERE ARE no words to tell you of the heat that consumes the Ganges plain during the months when the winter has left and the monsoon rains not yet begun. No shadow falls from the cloudless sky. Every breath is searing. A torpor settles over the villages strewn throughout the hard, stubbled fields. Children are subdued, adults languish on their charpoy beds, and even the cattle seem immobilized.

On one such burning day in April, in a village in Bihar, Susheela Devi was worried about her sick child. Susheela, a tired-looking woman in a tired-looking sari, is middle-aged at thirty-two years. She was given away in marriage when she was thirteen. Burdened by work, harassed by a bitter mother-in-law, and uncared for by a husband twenty years her senior. That crone mother-in-law still complained that Susheela had not brought sufficient dowry into the family. This was dangerous muttering. Other village wives had fatal "accidents" and were replaced with younger wives and new dowries. Her decision to seek medical help required brave determination. There was the overpowering heat through which she would have to walk, carrying her sick child most of the eight miles to the government health center. The peasant poor of India do not squander their precious rupees on the luxury of bus or bicycle rickshaw unless there is a great emergency. The child was ill, but not emergency-ill in any of the too familiar life-threatening ways . . . the acute fever and coma of childhood malaria, the rapid wasting diarrhea and death of cholera, or the labored gasps of pneumonia. It was merely that the child seemed somewhat feverish this past month and was becoming emaciated, despite a reasonably good appetite, with a distended abdomen like the children of the failed-monsoon famine years.

Susheela awoke before dawn the next morning; cooked the family's food for that day—some rice and a bit of curried river fish. She led the cow from the stall, which was attached to the house as a kind of guest room, and gave it some fodder. The cow was indeed an honored guest. The mount of Lord Siva as he rode through the cosmos and his audience when he danced, it now gave milk and *ghee* to this poor family. Its dried droppings were the chief source of fuel in the deforested plain. The very substance of the house's mud walls was the excrement from this holy beast.

The fitfully sleeping child was roused as the first light began to appear in the eastern sky. Mother and child left the village to begin the long walk to the health center. They were not long on the road when the sick child could walk no further and the 90-pound mother began to carry her child mile after painful mile. Three hours later, when the sun was rising above the neem trees, the exhausted Susheela arrived at the health center. It was about 9:00 a.m.; already more than one hundred people waited. They filled all the benches on the center's veranda. The overflow, joined by new arrivals, sat on the open ground surrounding the health center.

Susheela sat amongst the outside group. Many were

mothers with their sick children, but more were men. There were men with open wounds. Accidents from road and agriculture take as heavy a toll as malaria and other infectious illnesses. Women were but women; they had to be sicker than men to leave their work to go to the health center. The waiting patients were mostly silent, not unlike the withdrawn silence of our own physicians' waiting rooms. A half hour later the two doctors assigned to the health center arrived. They looked so young and, to Susheela, so different from the young men of her village. And they *were* different. They were born in the city, schooled in the city, medically educated in the city, and clinically trained in city hospitals. Upon graduating, they were forced to work a year or two years at a rural health facility. The shorter time (or no time) if their family was well connected. And as soon as they finished their country time they would be back in the city, scrambling against an excess of physicians to establish a practice. Only a few well-trained doctors chose to serve these non-paying country peasants. Their family would have to be conspicuously "unconnected" to be left in *that* limbo.

The hours passed. The heat intensified. Susheela and her child remained fixed to their bit of ground, afraid to leave even for water lest it be usurped by another of the ailing. Shortly before noon, a health center attendant who had been circulating amongst the waiting people, registering their names, came to Susheela and whispered that within the hour the doctor sahibs would leave for their lunch and would not return since in the afternoon they conducted their business seeing paying patients at their private clinics in the town. However, he confided, for 10 rupees (about 50 cents) he could arrange for her to see one of the doctor sahibs, the smart one, within the hour. Susheela was stunned by the demand, but not surprised; these privileges must always be bought from the petty government servants. From her small hidden reserve banked in a secret recess of the wall by her pallet she had withdrawn 7 rupees, half her "account," to take with her. She had decided to use it for bus or bicycle rickshaw to take her back home. She felt she just had no strength left to carry her child those eight miles. That would not now be possible, and she offered the attendant all she had, the 7 rupees. He grumbled and reluctantly agreed to accept the bribe, but for such a paltry sum she would not be able to see the smart young doctor sahib who had been at the health center for six months. She would have to be attended to by that ninny who had just graduated from medical school and had been posted to the center only two weeks ago. The precious few coins were handed over, a day's wages for many peasant laborers.

The attendant, fortunately, was an honest man and good to his bribe. Within a short time he called Susheela's name to enter the doctor's cubicle. The young doctor was brusque, unfriendly and uncommunicative. He was unsure of his skills but sure of his importance in being a doctor. He felt alien in this country setting and so he adopted the manners, experienced since childhood, in dealing with social inferiors—brusque, officious, and unfriendly.

He told Susheela to put her frightened child on the bare wooden examination table. This done, the doctor prodded the child's distended abdomen, his fingers sensing the greatly enlarged spleen and the enlarged liver whose boundary was well below the rib cage. The thermometer revealed the child's temperature to be 101°F. Without preliminary explanation he took the child's arm and swabbed it with an alcohol swab, which came away black with grime and sweat, and then, while the attendant took a tight grip on the thin arm, the doctor stuck a syringe needle into a vein. The procedure really didn't hurt that much, but for the frightened, apprehensive little girl, who had been to this moment so stoically uncomplaining, the sight of blood welling into the syringe barrel produced a shriek, followed by uncontrollable sobbing. A river of sad tears washed the small, pinched face. Susheela did the best she could to comfort her child and was told to wait in the corridor while the blood was sent to the laboratory to be tested.

The health center's laboratory was a small, dimly lit, not very clean room, cluttered with broken but still usable bits of glassware such as microscope slides and test tubes washed innumerable times, and some basic reagents to perform basic, all-important tests: blood counts, staining for malaria parasites, and the presence of abnormal sugar and protein in urine. There was also a microscope of Polish manufacture whose optics, not of good quality to begin with, had acquired a bloom of glass-loving fungus which gave a blurry, chiaroscuro image to the scrutinized malaria parasites, worm eggs, and other microscopic faunal and floral parasites.

The syringe was given to the laboratory technician, who expressed the blood into a glass centrifuge tube. On the floor in a corner of the laboratory, almost hidden in the gloom, sat an old man clothed only in a ragged dhoti. This peon was the centrifuge wallah, and it was now the moment for him to do his work. The health center had no electricity. The microscope was illuminated by the light that filtered through a shuttered window, and the centrifuge—a simple, motorless instrument—was propelled by means of a hand crank and gears. The centrifuge wallah cranked vigorously, spinning the blood tube some 1500 revolutions each minute. After five minutes, the cranking stopped and the machine came to rest. The centrifuge wallah returned, wordlessly, to his place in the corner.

In the tube, the centrifugal force of spinning had separated the blood into its components, a bottom layer of packed red cells above which there was a thin stratum of white cells. Over the packed sediment of cells was the straw-colored fluid—the serum. In normal, healthy people, there should be an almost 1:1 ratio between the volumes of packed red blood cells and the serum. But in this specimen, the technician noted that the tube contained only about 25 percent in packed cells. The child had only about half the red blood cells that was normal . . . a case of severe anemia. Then the technician transferred a few milliliters of the serum to another tube. To that tube he added a few drops of formaldehyde and vigorously shook the mixture for a few seconds. Within minutes the serum began to solidify into a Jell-O-like clot. This was an abnormal reaction that occurred only when a very great amount of gamma globulin was present in the serum.

To the doctor, the constellation of signs and symptoms could point to only one diagnosis. The prolonged fever, the greatly enlarged liver and spleen, the anemia, the serum that gelled when mixed with formaldehyde, all meant visceral leishmaniasis, a disease that both peasant and physician knew by its vernacular Mogul name of *kala azar*—the "black sickness." Realizing the gravity of what he was about to tell Susheela, his pomposity fell away. The young man had not yet hardened to his role as death's messenger. "Mother," he said gently, "your child is very ill with kala azar." The words made Susheela gasp and her eyes stung as if she had been struck with great physical force. The people of her village and the other villages of the district were no strangers to kala azar. The disease had reappeared some ten years ago, to begin killing old and young alike, but mostly the young. Before that time, her generation was ignorant of the disease, although the old men of the village spoke of former times, during the days of the British Raj, when great epidemics of kala azar ravaged Bengal and Bihar, the Terai of neighboring Nepal, decimating villages, towns, and cities. Susheela's child was a new victim of the new epidemic.

With urgency, the doctor continued: "It does not mean death. Your child can be cured. You must buy medicine. Then you must come here every day for twenty days so the nurse can inject the medicine. Every day! Not a single day must be missed!"

The word that echoed in Susheela's head was "buy." *Buy?* Buy medicine? She had brought her child to be treated, not to be given a few words of advice. She had walked those miles, sat for hours waiting in the sun; she had paid her bribe—every rupee she had—to see the doctor. Now she was told to buy the medicine that she expected the health facility to provide her.

"Cannot I get the medicine from you, doctor?"

"No," he replied. "We have no kala azar medicine. The state government in Patna gives our dispensary only a few simple medicines to treat a few simple illnesses. Last year we had some kala azar medicine, but there were many people to be treated and our supply was soon finished. We've written again and again to the central medical stores in Patna for more drugs, but they do not even reply to our letters. Now everyone must buy the medicine where they can and as best they can."

"What will it cost?"

"That I cannot tell you. If the pharmacist has a good

stock, it will be cheaper; if not, it will be expensive." He scribbled a prescription on a scrap of paper and, reverting to his customary brusqueness, thrust it at Susheela, peremptorily dismissing her with "Unless you get the medicine there is nothing more I can do for you. Go now!"

Susheela made her way through the crowded market until she found the druggist's shop. She handed him the scrap of paper on which the doctor's prescription was written and the druggist told her that she was indeed fortunate because there were only two bottles of the kala azar medicine left. Many people in the district had the disease and needed the medicine.

"How much is it?" Susheela asked fearfully.

"Mother," replied the druggist, "I know that you are poor and your child is ill. For you, I will give you a bottle of the drug, enough for her whole treatment, for three hundred rupees [about $15]. Others I would charge five hundred rupees."

Three hundred or 500 rupees; it made no difference. It was an astronomical sum, more than the family's income for some months. Susheela picked up her child, turned from the pharmacist, and began the long walk back to her village.

It was well after dark when she reached home. There is no purpose in recounting the abuse she received from her mother-in-law and her husband for being away from her duties for an entire day . . . that was too customary. Later, as they lay on their sleeping mat in the darkness, Susheela told her husband what had happened that day. They must buy the medicine. But how? Sell the cow? If they did that, then surely the family would starve. Borrow from the landowner from whom they leased their small plot of rice paddy? Impossible. They were already in debt to him for almost half the future crop to pay the lease rent and to repay the loan he had advanced to buy seed and fertilizer. There were no family resources. Each relative was as poor and indebted as themselves. Even if by some miracle they could buy the medicine, there was no way that Susheela and the child could travel those long miles to the health center for twenty consecutive days. Not only, she knew, did she lack the strength but the monsoon rains would soon begin and it was essential that she help prepare the soil and plant the rice. The family's precarious survival required the labor of each member at planting and harvest. No, for the child they would have to do the best they could. They would pray to the gods. They would consult the "doctor" in tne adjoining village, a man who practiced Ayurvedic herbal medicine.

In the weeks that followed they prayed at the small village shrine, leaving offerings of food, as if the gods too were petty bureaucrats who had to be bribed. The Ayurvedic doctor did not miss the diagnosis of kala azar and for a few rupees gave the child some herbal medicine that for over a thousand years had been prescribed for fevers. And indeed, for a short few days after taking the draught, the child's fever abated and she brightened. But in the end, as the weeks passed, she became progressively more ill: she grew even more emaciated, her skin turned a dusky gray, her hair became brittle, small bleeding sores covered her body, and the abdomen, burdened with a grossly enlarged liver, distended even further.

One day, some three months after Susheela's visit to the health center, the child began to cough and gasp for breath. During the night, the little girl died.

The mourning family carried the small body on a wooden plank, draped with a bit of cloth and adorned with marigolds, to the banks of the Ganges. There at the burning ghats the body was offered to the fires. A fragment of life sacrificed for want of $15.

For the family there was some comfort. There were seven other living children. And, by good fortune, it was not a boy that had died.

Chapter 3

How the Government Disease Came to India

THE INSTRUMENT WHICH consigned Susheela's daughter to the funeral pyre was a tiny midge, no weightier than an eyelash. In Patna, the capital of Bihar State, fifty miles from Susheela's village, Dr. A. K. Chakravarty was holding a cage in which these silver-winged insects were quietly resting on the screened walls. "They look so innocent" was his brooding observation on the treachery of appearances. Chakravarty is the chief of India's National Institute of Communicable Diseases' Kala Azar Unit, based in Patna. He is a large man, well over 6 feet tall, and despite his determined, rather ferocious countenance, he is of gentle nature: a devout Hindu, given to philosophic enquiry into human spirituality; veterinarian-researcher (more later of why a veterinarian should be in charge of a unit conducting studies on a human disease) who treats his animal and human patients with a delicate, kindly courtesy. The objects of this disproportionate encounter between man and insect were the blood-sucking sandfly midges, *Phlebotomus argentipes*. As the anopheline mosquito is the biologically required transmitter of human malaria, the phlebotomine sandfly is the purveyor of another killing parasitic protozoan, *Leishmania donovani*, the cause of kala azar.

Of all the parasites, great and small, that make our bodies home, the *Leishmania* may well be the most peculiar and intractable. To the invaded host, a parasite is a foreign body calling for a response. A major effector mechanism of immune defense is the killing and devouring of the parasite by specialized wandering and fixed cells, the front-line soldiers of the immune system, known as phagocytes. The *Leishmania* have the effrontery not only to evade digestion by the phagocytes but actually to invade them in an obligate fashion to become intracellular parasites.

The parasite was but one factor that led to the funeral pyre of Susheela's child. Her tragedy was an interweaving of parasite and its sandfly transmitter; climate and culture; society, medicine, and politics. Let me take up the thread of this complex of interrelationships by examining the history of the *Leishmania* and the disease it causes—kala azar.

To understand the history of kala azar (or any other human disease), we must keep in mind that our "files" go back only a few thousand years . . . to the beginning of written language, the beginning of the human documentary. There are a few mummies and ancestral bones from which the medical archeology detectives have gained sparse clues into the nature and epidemiology of ancient illnesses. However, humans must have thought about their health from that imperceptible time when they became truly human—an animal with the cognitive sense of well-being, and of illness and the certainty of death; an animal with the gift of foresight. Health and disease was an issue to our ancestors as it is to us today. When the ancients devised the written word, they wrote about their illnesses—in hieroglyphics—in Sumerian, Babylonian, Arabic, Greek, Latin.

All early written languages seem to have had their

"medical books," although notions of disease causation were very different from ours. Microbes and such are new to the medical profession. Before the first microbe could even be discovered there had to be a good microscope to see them by, and not until about 1825 were achromatic microscope lenses available. Then someone with extraordinary insight had to put two and two together to recognize, for the first time, that so minuscule an organism could be the specific agent of a specific disease. Only in 1875 was a protozoan shown to be a pathogen. This was *Endamoeba histolytica*, the cause of amoebic dysentery and amoebic liver abscess. This "tropical" parasite was described by F. Lösch from Russian patients living in the balmy latitudes approximately a hundred miles south of the Arctic Circle. Viruses are newer still. The first virus to be discovered was a virus not of humans but of plants. It was the tobacco mosaic virus described by the Russian D. Iwanowski, in 1892. The yellow fever virus, discovered in 1900 by the U.S. Army Commission led by Major Walter Reed, was the first viral disease of humans to be identified.

While the pre-Pasteur doctors may have had some peculiar ideas on the causation of disease—demons and devils, humors and miasmas—they weren't all that bad as descriptive clinicians. They recorded the signs and symptoms of their patients' complaints with sufficient accuracy to allow us to identify some of those illnesses in modern terms. Malaria, for example, with its cyclical, regular periodicity of rigor-shakes and fever-sweats, was recognized as malaria by physicians of ancient and medieval Europe, the Middle East, and Asia a thousand or more years before November 6, 1880—the day that Alphonse Laveran, a French Army médecin-major 1re classe, posted to Algeria, saw under his microscope the malaria parasite within the blood cells of a feverish, twenty-four-year-old artillery man. Thus we can look back with reasonable confidence and make historical-epidemiological judgments, saying that malaria was the scourge of ancient Rome and that it has persisted in southern China for at least two thousand years.

Kala azar, however, is different: it has not yielded its past origins and epidemiology to the modern historian's search. This is peculiar because the disease is distinct enough that had it been present it should have been descriptively remarked upon in the early medical writings. It is not an indolent disease that would have gone unnoticed. Kala azar frequently occurs in epidemic proportion, killing thousands during its apogee. Nor can we attribute this narrative absence to an observational gap solely on the part of India's ancient writers. Visceral leishmaniasis (kala azar) is not confined to India but is now known to occur in a vast area of China, in Russian Turkestan, in the Sudan and Ethiopia, in Mediterranean Europe (southern Spain, France, and Italy, Greece, Malta, Crete, and Yugoslavia),[2] in North Africa, and in the New World as foci of infection along the coast of Brazil. Except for Brazil and the Sudan, these are regions with a rich written record spanning at least fifteen hundred years. In that record we can see the past epidemics of plague, typhus, malaria—but nowhere do we find an account of a disease that could be interpreted as kala azar. To the best of our admittedly imperfect knowledge, kala azar seems to have made its first attack on humans in Jessore in 1824. It was like a new scourge, as AIDS is in our time. And like

2. The tourist and guide books certainly do not mention the risk of contracting kala azar at these European tourist meccas. The odds are admittedly very small. Still, a tourist stands a greater chance of getting kala azar in the French Côte d'Azur than breaking the bank at Monte Carlo. And while it is hardly a tourist resort, special and timely note should be made of kala azar's entrenched endemicity in Iraq.

AIDS, its true epidemiological origins may never be satisfactorily traced.

In recent times, Jessore has been a pawn of political events. It is now in Bangladesh, but its spirit defies national boundaries; it remains a city of Bengal. The Bengalis of Jessore may worship Allah, yet they recite the poems of Tagore with the same passion as do their Hindu cousins a short march to the west in Calcutta. Now a sleepy market city near the Indian border, Jessore was an important commercial center during the Mogul Empire. From the mid-1700s it was successively administered by the East India Company and then by the British government itself. Jessore probably hasn't changed that much since the first British East India Company agent-administrators, appropriately known as Collectors, were posted there in the 1750s—except that in the 1750s there weren't any Chinese restaurants, a favorite cuisine of modern middle-class Bangladeshis. Many of the Collectors were honest men of noble intent who had real concern to alleviate the lot of the tax-oppressed peasants. One Collector of Jessore, with the unlikely name of Telman Henkel, was so popular that the locals made an effigy of him and worshipped it. That doesn't happen to Indian political administrators today. The Collectors would certainly have noted and reported in their journals the presence of a killing epidemic of the dimensions that kala azar was about to become. Moreover, by 1764 the Collectors were joined by civil and military surgeons of the newly established Indian Medical Service in their district.[3]

It began in the village of Mohamedpur, thirty miles dis-

3. Hospitals were established very early in the East India Company's rule: in 1664 in Madras, 1676 in Bombay, and 1707 in Calcutta. The British doctors in the Company's employ earned the princely annual wage of £36.

tant from Jessore. In the last months of 1824 the people of Mohamedpur began to die. Their color darkened to a clayey gray. The flesh fell away and the abdominal veins stood out, like enlarged blue cords, on wasted bodies. An overwhelming dysentery or pneumonia were the common final events that terminated their lives. With frightening progression the "black sickness" engulfed Jessore District and then enveloped the entire Gangetic plain. By 1832 it had spread from Jessore into what is now Indian West Bengal. The disease progressed by the arteries of road and water as if it were a systemic infection of a dying land.

Even before the Mogul conquests, the capital of Bangladesh (then Dacca, now Dhaka) was a thriving river port city. A network of rivers from upcountry feed into the great Jamuna (the lower Brahmaputra) in the Dhaka District. Lateen-rigged sailing barges with brilliantly colored sails and pole-propelled craft of all sizes on which whole families and crews live under cramped thatch cover continue to crowd the riverbank at Dhaka, bringing produce and manufacture from all over Bengal-Bangladesh. Although it is some two hundred miles from the Bay of Bengal's Ganges estuary, in the Jamuna that flows by Dhaka and its environs one can see the river porpoises, giving the appearance of monstrous, fabled sea serpents as their bodies roll in the water.

In 1862, one of these boats from upcountry brought a consignment of rice to Jageer, a populous town near Dacca. For over six months the entire crew had been ill with an intermittent fever and lassitude. Economic necessity had required them to pole-push and tow to Jageer, the men straining along the riverbank's towpath at the length of rope attached to their heavily ladened boat. It was to be their last trip. In Jageer, their condition rapidly deteriorated, and one by one they died. It is believed that these boatmen from upcountry were the "spores" that introduced kala azar to the Dacca District. Over the next four years the mortality in Jageer was incalculable. Perhaps only descriptions of bubonic plague epidemics compare. The dead lay where they died, abandoned in their homes, or were thrown into the rivers or *beels* (the artificial irrigation ponds in most Bengal villages). Four years later, Jageer as a living community had ceased to exist. Today it is no longer on the map, and the curious traveler cannot exactly locate its remnants.

In 1876 an Indian physician, Dr. G. C. Roy, published an account of that time. His words, written before the discovery of the cause and transmission of the disease, remain a model of clinical-epidemiological observation: "The mode of attack of the villages one after another is very peculiar. In the first year, the villages adjacent to an epidemic-stricken locality will show at the close of the rains more of ordinary fever cases and greater mortality than usual, but this being nothing more than they are accustomed to in some fever seasons, will not create any alarm or grave apprehension." Roy then notes that with winter the number of cases declines and "the people congratulate themselves on the change." The full brunt of the epidemic begins with the second year's rainy season. The disease "becomes more general and the village is panic-stricken. Deaths from acute fever run high. The suffering being general, there is seldom any person spared in a family to attend to the sick."

Roy's observations of despair are confirmed in the journal of a British civil surgeon of that time, a Dr. French. He writes of whole villages "in which not a healthy person was to be met with, while repeated relapses of fever, the daily deaths, the loss of their children, the increasing depopulation of their village, the absence of all hope for better times, had so demoralized the population that they neglected to avail themselves of medical and other aid, unless brought actually to their homes."

The British make no apology for their imperial period. They speak with pride of their high purpose as colonial custodians in freeing the peasantry from the excesses of despotic native rulers; in endowing their former colonies with a judiciary, a sense of fair play, cricket, and a democratic government. The French, on the other hand, endowed their colonies with the ability to bake wonderful loaves of bread. Making a current comparison between the former colonies of those two powers it would often seem that good bread has proved to be more sustaining and enduring than hand-me-down parliaments. Less often mentioned, and of equal importance to the ultimate character of these colonies-become-nations, was the fervor with which the British built the avenues of communication—roads, railways, and waterways. This was particularly true in India where, in the first half of the nineteenth century, they built the Bombay-Agra Road, the Bombay-Calcutta Road, and the Grand Trunk Road from Calcutta to Peshawar. Three thousand miles of new roads—all paved. They also built a canal-irrigation system for the Ganges and its tributaries that when finished was the most extensive in the world.

The Brits were not the Mad Ludwigs of road building, nor were their motives purely altruistic. Commercial prosperity (or "exploitation," if you consider colonialism to be a four-letter word) demanded the means to collect and distribute the produce of the country and the manufactured goods of the motherland. Commerce also demanded a pacified dominion. The pacified dominion, the Pax Britannica, required the roads, the railways and waterways to deploy the troops and administrative officers where needed.

Unfortunately, what was good for the business of colonial rule was also good for the pathogens. The new corridors that brought the rice and lentils, the Manchester bolts of cotton cloth and cooking pots, were also passageways for the dissemination of infectious disease.[4]

Kala azar was a stowaway traveler to Assam, carried there in 1875 by the British steamers that began to ply the upper Ganges and Brahmaputra rivers. When the infection broke out in Assam, the inhabitants recognized it as something new to their experience and somehow associated the disease with the activities of their new masters, the British.[5] With remarkable epidemiological insight, they called their new affliction *sakari bemari*, "the government disease." Now Assam was ignited, and during the next twenty-five years kala azar in some districts killed 25 percent of the population. Some villages lost two thirds, or more, of their people. From Assam to Tamil Nadhu, kala azar had established a permanent residency in India.

4. An account of the untoward effects of road building on disease could fill a book. Not only do roads facilitate the dissemination of infection, as was the case of the kala azar epidemic in India and of sleeping sickness in tropical Africa, but the ecological consequences of road building frequently are the creation of habitats favorable for the breeding of insect vectors. The breeding of malaria-transmitting anopheline mosquitoes in the collections of water on rutted roads and their verges is but one example.

5. Assam, as much a part of Burma as India, was not originally included in Britain's Indian "package." This wild hilly country of thick steamy jungles inhabited by fierce naked tribes fell to the British as a spoil of war with Burma in 1826.

Chapter 4

In Search of Kala Azar: Bedbugs and Other Red Herrings

WHEN THE CENTURY turned to 1900 the epidemic in the Gangetic plain began to wane.[6] After kala azar's merciless half-century hold in Bengal, Bihar, and Assam, the demoralized and depopulated land slowly began to return to a more secure and prosperous life. Kala azar did not completely disappear; cases continued to occur, but at much lower, non-threatening numbers. However, when the epidemic ended, an inquiry could be pursued that was impossible when the disease first struck Jessore in 1824. Between 1824 and 1900 the concept of the causation of disease had undergone a radical change. It was as if medical science, during this brief period of time, had emerged from the long medieval night into the enlightenment. Louis Pas-

6. Most epidemics of infectious disease seem to have a natural cycle of activity. After various lengths of time there is a decline in intensity even when there is no apparent human intervention or alteration in ecological or behavioral factors. The mechanisms that govern the rise and fall of these epidemics are still imperfectly understood. Mutational changes in the pathogen to a less virulent form, acquisition of a herd immunity on the part of the populace, or, simply, that so many people have died that the pathogen can't get around much anymore, are all possible explanations. But mostly, as in the case of the 1824–1900 kala azar epidemic in India, it is the X factor . . . the unknown. Kala azar seems to have an epidemic cycle of fifteen to twenty years.

teur ushered in the new science with his studies on the microbial causation of "sick" beer and silkworms. By the 1870s he was extending his studies to the microbial pathogens of animals, and then to those of humans. It was not long thereafter that the Germans entered the game, led by the giant intellect of Robert Koch. Now microbial pathogens were being discovered almost monthly—cholera, plague, boils, diphtheria. It seemed then that microbes were everywhere; that all disease had a microbial etiology. There was an enormous excitement and vigor in the pursuit of the pathogen. The *Zeitgeist* of the period was expressed by old Dr. Gottlieb in Sinclair Lewis's novel *Arrowsmith*, who "anoints" the young scientist with the benediction, "May Koch bless you!" Parasitic pathogens were also being discovered during those years; Laveran (as noted earlier) unearthed the malaria parasite in 1880, and J. E. Dutton found the trypanosome in a human sleeping sickness patient in 1902. Science and tropical medicine were being brought together. Paul Ehrlich's work in Germany on dyes and drugs began the search and discovery of synthetically created, powerful chemotherapeutic agents. Diseases were beginning to yield up their secrets *and* they were yielding to cure by man-made drugs.

The diseases of the tropics were studied by scientist-physicians who were truly *engagé*. Some of the best minds that the ruling country had to offer went to the colonies. They watched birds, catalogued the fauna and flora, sat up nights in flimsy *machans* to kill tiger. They played polo, shot snipe—and pursued their microbial quarry with a tenacity that by today's standards of paid-for-by-project research seems almost quixotic. So, when the old Queen went to her reward in 1900, there was already in place in India a medical research establishment staffed mainly by military men of the Indian Medical Service. Other excellent research was carried out by medical men, such as tea estate physicians, working in remote areas under primitive conditions. It was at this time and in this climate that these men (and one woman) took off in hot pursuit to discover the causative agent of kala azar. It was a pursuit that would follow many false trails and scents before the *Leishmania* was bearded in its macrophage and sandfly dens.

It seems strange from today's retrospect that the first false trail was laid down by a worm—the hookworm.[7] The ancients were ignorant of the hookworm but aware of the disease it caused. Almost one thousand years ago a Chinese medical commentary epitomized it as the "able to eat but too lazy to work" disease. In 1838, a Milanese physician, Angelo Dubini, described the numerous small worms that he saw attached to the intestine of a dead Italian peasant whom he had autopsied. He saw the worms but didn't understand what they could do. That was left to O. Wucherer, a physician born in Portugal of German descent and practicing in Brazil. In 1861 the Benedictine

7. There are two hookworms of humans, *Ancylostoma duodenale* and *Necator americanus*. The adult worms are attached by fanglike mouthparts to the inner wall of the small intestine, where they assiduously suck the blood from small, ulcerous lesions at their attachment site. In time, with enough worms (and often compounded by other conditions such as iron-poor diet, malaria, and pregnancy), the constant loss of blood leads to a severe anemia—even death in the worst cases. Hookworms are both temperate and tropical in their geographical distribution. At one time they almost bled the American South white. The hookworms also caused great debility to miners because the then unsanitary conditions in mines and tunnels were highly favorable for hookworm transmission. Hookworm eggs pass out with the feces onto the soil and hatch into larvae which remain quiescent until trod on by the unshod human foot. They then penetrate the skin of the foot and make their way through the body to the intestine, where they grow to adult size, mate, and suck blood. In the tropics the barefoot boy is just as likely to be palely anemic as brown-cheeked.

monks of Bahia, with true Christian charity, called Dr. Wucherer to attend to a dying black slave they owned. The wretched man was at death's door, so severely anemic that his blood was almost water. The next day the slave died, and over the strong objections of the monks, Wucherer did a postmortem examination and found masses of "Dubini's worms" clamped to the intestinal wall. From this he made the association between the parasite and the "hypoaemia." Thus, in 1890, when a commission was sent to Assam to investigate the cause of kala azar, they searched for the then known agents that could cause severe anemia, and they thought of hookworm. Hookworm was a natural prime suspect. And when the investigator in charge, Dr. Giles, examined the feces of inhabitants of kala azar-stricken villages, he did indeed find the characteristic thin-shelled eggs of the hookworm under his microscope. "Kala azar is hookworm," said Dr. Giles.

"Not so," said Surgeon-Major Dobson, also posted to Assam. Dobson's choice was that malaria caused kala azar. It was conceded that hookworm would cause anemia, but in kala azar not only was there an anemia but there was also a big spleen, and splenic enlargement was not a consequence of hookworm disease. Malaria caused anemia *and* splenomegaly. Kala azar was malaria.

"Not so," said Dr. Giles. The big spleen didn't count. Practically everyone in Assam had or had had malaria. There were a lot of people without kala azar walking around with a big spleen. Besides, the fever of kala azar was not like that of malaria, it was more sustained and unremitting, and there was that progressive downhill slide to death that was atypical of malaria.

The malariaphiles (by 1896, Dobson was joined in his views by others) continued the forensic with the rebuttal that practically everyone in Assam had hookworm but not everyone had kala azar. Kala azar was a *special* form of malaria—a malaria cachexia (a progressive wasting condition). Meanwhile, wiser heads maintained that kala azar's cause was neither hookworm nor malaria but a pathogen yet to be discovered. When that pathogen *was* discovered in 1900, it turned out to be a microbial organism, a pathogenic protozoan, entirely new to human knowledge.

If you have to go to Calcutta, and if you have to fly there, you will land at Dum Dum Airport. It is a terrible place. Noisy, unbelievably crowded, unbelievably confused, and retrieving your luggage (if it beats the odds and actually arrives on your flight) is an exercise exemplifying Darwin's principle of the survival of the fittest (and the pushiest). One hundred years ago it was also a terrible place. The town of Dum Dum and the British cantonment there (about ten miles from Calcutta) was so beset by kala azar that the vernacular name for kala azar throughout Bengal was "Dum Dum fever." In 1900 one of those stricken was an Irish British soldier from the military cantonment. There were no drugs then for kala azar, although he may have been given quinine in the mistaken belief in the malaria cachexia hypothesis. The soldier was invalided to the military hospital in Netley, England, where he died. His body was autopsied by a Dr. William Boog Leishman, formerly of the India Medical Service. Leishman—bald, beaky-nosed, military moustached (for some reason I think him a handsome man)—was a physician possessed of the spirit of scientific inquiry present in so many others of his IMS colleagues. He was also on the hunt for the causative organism of kala azar. Leishman excised a bit of tissue from the dead man's grossly enlarged spleen, stained the samples (with a stain later called Leishman's stain; we still use it to examine blood for malaria and other blood parasites), and examined the preparation under his brass-barreled microscope. There under the microscope lens he saw the numerous, ovoid-shaped forms within the macrophage cells that populate the spleen. The sporelike bodies were minute, not much bigger than a bacterium.

This British soldier, nameless in medical archives, unwittingly gave his life for science. His body yielded up the secret of his last enemy. But what was it? Now began the biological-taxonomic pursuit that was to occupy tropical medicine specialists, parasitologists, and entomologists for the next thirty years. Even though the causative organism had now been observed, false trails continued to lie ahead. Leishman had once seen trypanosomes[8] in the blood of an Indian rat, and although his kala azar organisms were intracellular and much smaller than the trypanosomes, there were some similarities in structure. Putting two and two together (which, unfortunately, turned out to add up taxonomically to three), Leishman erroneously concluded that kala azar was caused by trypanosomes, and that the "bodies" within the splenic macrophages were actually degenerate forms of trypanosomes that had been ingested (phagocytized) and partially digested by the host cells.

Now others began finding the "Leishman bodies," the first confirmer being Charles Donovan in Madras. Unlike Leishman, Donovan studied the sick rather than the dead. He began a diagnostic practice that is still used, of sticking quite a large needle through the patient's skin, through the

8. Trypanosomes are the cause of African sleeping sickness. Other species of trypanosome parasitize a wide variety of other animals from frogs to primates. They are microscopic "fishlike" protozoan organisms swimming in the bloodstream by means of a lashing flagellum. They belong to the larger group of hemoflagellates, and *Leishmania* are of this group—first or maybe second cousins to the trypanosomes.

abdominal wall, and into the body of the spleen. Donovan expressed the plug of splenic tissue captured in the needle onto a glass slide, stained it, and examined it under the microscope. The organisms which Leishman had described were seen only in the splenic tissue of patients with symptoms typical of kala azar.[9] By 1904 the organisms were recognized as being protozoan in nature. They were given the name "Leishman-Donovan bodies," and later the taxonomic designation of *Leishmania donovani*.

But nomenclature is not an end in itself. The tidy housekeeping of biology demands that any newly discovered creature or plant be classified according to its affiliations—its family ties—siblings, near and distant cousins, the near and distant relatives of its extended family. In more exact terms, it would need a sorting placement within genus, family, class, and phylum. In 1903, *Leishmania donovani* was still a taxonomic orphan in search of its nearest relatives. Of more crucial importance, however, was the unsolved question of how *Leishmania donovani* got from A to B . . . from an infected person to its next host. The disease could not be effectively controlled unless the manner of its transmission was known. The spread of kala azar from household to household, and from village to nearby village, clearly indicated that it was an infectious disease. Was it carried in the air from the breath of the infected to the uninfected? Was it carried in the drinking water contami-

9. The discoveries made in India were the "main event" of the investigations on kala azar, and what happened, and is happening, in the Indian subcontinent is the theme of my story. However, as noted earlier, kala azar (visceral leishmaniasis) occurs elsewhere too and early research was carried out in those endemic regions. In 1903, Marchand described the parasites from a British soldier who died during the fighting near Peking. That same year Pianese found the organisms in smears from the spleen and liver of children dying of "splenic anemia" in southern Italy.

nated by the excretions of the sick? By touch? By sex? By giving nursing care to the stricken? Or by another route that in 1903 was beginning to appear as a new possible mode of transmission—the blood-sucking insect?

In 1876, (Patrick Manson) who was to become the father of tropical medicine), working in Amoy, China, as a physician to the Chinese Imperial Customs Service, discovered that the filarial worm, the cause of elephantiasis,[10] was transmitted by the mosquito. Then, in 1898, Ronald Ross in India and G. B. Grassi in Italy made the great discovery that malarial parasites were also transmitted by mosquitoes.[11] So, by analogy, there were those who hypothesized a blood-sucking insect as the means by which *Leishmania donovani* got from A to B.

One of the first clues to both transmission and taxonomy came from the "test tube" cultivation of the parasite. It is all very routine today: you go to your doctor with your sore throat or other infected part, and a sample is taken for cultivation on artificial medium, the organism grown out, identified, and its sensitivity to various antibiotics determined so as to select the most efficacious one for your treatment. In 1903, when microbiology was in its lusty infancy, cultivation techniques were just being devised. It was essential to isolate, propagate, and identify the microbe in culture for research on the origin and treatment of infec-

10. It is said that Manson's first encounter with elephantiasis was a Chinese peanut vendor whose scrotum was so large that he used it as a counter to display his merchandise.

11. That a parasitic organism living in a warm-blooded host will undergo a profound morphological and physiological transformation to complete its cycle in the completely alien milieu of the invertebrate is a truly remarkable adaptation. We still have no real understanding of the cues and the consequent genetic activation that bring about these transformations.

tious diseases. In each case, the discovery of a new pathogen was quickly followed by an attempt to grow it under laboratory conditions. Sometimes it worked. Sometimes it didn't. It was not, for example, until some seventy-five years after the discovery of the malaria parasite that the trick of getting it to grow in the "test tube" was worked out. For some microorganisms and parasites the culture trick has still not been elucidated.

Thus, when *Leishmania donovani* was identified as the cause of kala azar, the next natural step was to grow the organisms in man-made culture medium. In 1904, a year after Donovan's confirmation of the causative organism of kala azar, Sir Leonard Rogers, working in Calcutta, put some spleen tissue from a patient into a simple salt solution nutriment "soup," Sir Leonard must have been dumbstruck with surprise when, a week later, he peered down his microscope at a drop of the culture fluid. What he would have expected, if the culture was successful, was a multiplication of the Leishman-Donovan bodies, like so:

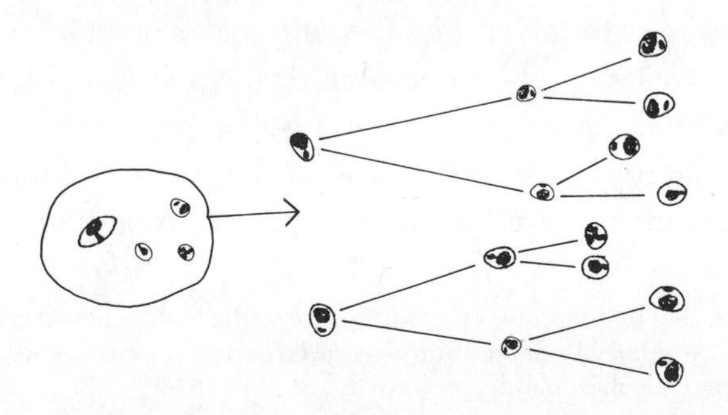

But that was not what Sir Leonard saw. The Leishman-Donovan bodies had *transformed.* From a minute ovoid form, they had become, in culture, spindle-shaped bodies some ten times the size of their midget Leishmanial parent, each bearing a single threadlike flagellum at the anterior end. These flagellated forms were multiplying by an asexual division in the culture fluid, like so:

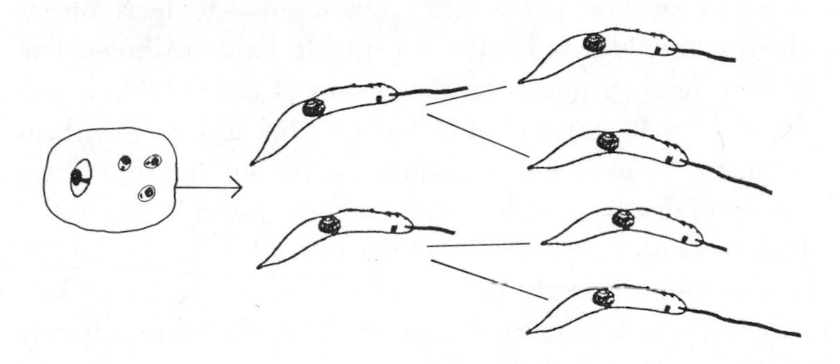

This meant that the Leishman-Donovan bodies within the macrophages of kala azar patients were but one stage in the life cycle of the parasite. Culture had revealed that there was a flagellated stage also, and this made it a cousin of the trypanosome, relatives within the larger family of tissue and blood-dwelling flagellates collectively known as the hemoflagellates. Protozoologists had seen the same spindle-shaped flagellate organisms (we now call this stage/form the *promastigote*) elsewhere—within the intestinal tract of a variety of flies and bugs. It was therefore most likely that what happened in culture represented what would normally occur in the gut of a blood-sucking insect. Ergo! Transmission of *Leishmania donovani* from person to person was by the bite of an insect infected with these flagellate forms. A good and sound assumption; but culture provided no clue as to what the insect—or insects—might be. Another thirty years of intense research were to pass and many more false trails laid before the innocent flies of Dr. Chakravarty were proven to be the guilty party.

The bedbug, even when dignified by its Latin name of *Cimex lectularius*, is a loathsome creature. During the depth of night it creeps from its hiding places—whether the cracks of mud-walled hovels in the tropics or the steam pipes and crannies of North America's tenements—to feed on its sleeping blood supply. It's not nice to have bedbugs. The scarred feeding marks on the bodies of poor children will attest to that. Logic would have it that any bug with as disgusting, blood-sucking habits as the bedbug must be a carrier of disease-causing microorganisms. And, God knows, there was no shortage of bedbugs in India. Thus, the bedbug became an early prime suspect as the vector of kala azar.[12] Those who incriminated the bedbugs became highly partisan in their belief and there was, for some twenty-five years, the tantalizing but not quite conclusive experimental evidence that was to keep them steadfast to the bedbug red herring.

The chief champion of the bedbug-as-vector was Dr. W. S. Patton, who was conducting his researches in Madras. For five years, from 1907 to 1912, Patton fed bedbugs on kala azar patients. In the intestine of a few of the bedbugs that he dissected several days after their blood meal, he observed the flagellate "culture" form of the parasite. However, although the Leishman-Donovan bodies from the patients had transformed in the insect's intestine, the flag-

12. Over the years, the bedbug has been suspected, incriminated, and condemned as being the transmitter of a variety of parasites, bacteria, and viruses, but has always been proven innocent. As far as is now known, the bedbug's bite does not transmit any infectious pathogens to humans. Lice were also an early suspect as the vector of *Leishmania donovani* but it was pointed out that the Bengali was clean of his person and rarely lousy. The bedbug, on the other hand, would bite the sleeping washed and the sleeping unwashed without partiality.

ellate-stage parasites could not be found in its salivary glands or mouthparts (by analogy, the infective form of the malaria parasite is in the mosquito's salivary glands). Those glands would have to be the repository of the parasite if they were to be transmitted during the feeding process.[13] Well, reasoned Patton, et al., if they don't spit *in* you, they shit *on* you—transmission was by the infective route. The flagellate bodies were voided in the insect's feces and made their way into the human body by abrasions on the skin or the small puncture made by the feeding bedbug.[14]

Meanwhile, the bedbug was giving researchers a considerable amount of trouble. The flagellates tended to disappear from the bug's gut after a few days and those organisms that did remain were mostly dead and dying forms. As for the feces, it was difficult to recover viable forms that would "return to life" in test tube culture medium. Not a candidate vector to bet one's career on. Then, in 1922, the bedbug was dramatically rescued by a Mrs. Helen Adie. Mrs. Adie was a protozoologist working on kala azar in

13. Salivary juice, "bug spit," is essential to the blood-sucking insect. It contains powerful anticoagulants (which causes the itch and to which one can become allergically sensitized) that prevent the blood from clotting, and clogging in the small-bored "hypodermic-needle" mouthpart.

14. By analogy, this was not a totally unreasonable hypothesis. There is an allied disease present in the American tropics, Chagas' disease, caused by *Trypanosoma cruzi*. It's an untreatable infection that affects hundreds of thousands of people and may lead to death by heart failure or other cardiac abnormalities. It is transmitted by another blood-sucking bug, the Triatomid, which can slip its stiletto mouthpart into the sleeping person so artfully that its familiar names are the "kissing bug" and the "assassin bug." The infective-stage forms of *Trypanosoma cruzi* are in the bug's feces. When the Triatomid feeds, it takes in blood at one end and simultaneously defecates on the skin from the other end. The sleeping host may unconsciously scratch the bite and rub the parasites into the body through a lesion to effect transmission of the infection.

Calcutta[15] who claimed to have actually found the *Leishmania* in the bedbug's salivary glands. This would be almost positive proof that *Cimex lectularius* was the One True Transmitter. It was summer and the government and health officials, anybody who was anybody, had moved to the hill stations. Mrs. Adie sent a telegram to the government sitting in Simla telling of her great news, and a few weeks later that telegram was published, *in toto,* in the *Indian Journal of Medical Research.* Adie's finding greatly buoyed Patton's cause. Later that year, he announced to the Indian Science Congress that the bedbug theory was now nearly complete.

In the meantime, Mrs. Adie's microscopical preparations of the infected bedbug salivary glands were sent to other experts and their judgment caused the rapid deflation of the bedbug-as-vector theory. The organisms in the glands were not *Leishmania* but rather a completely unrelated protozoan parasite, *Nosema,* that had a superficial morphological resemblance to it. *Nosema* was known to be a common parasite of insects and even played a role in Pasteur's formulating the germ theory of disease from his finding the protozoan in sick silk worms. Even after all these years I can feel so sorry for Mrs. Adie . . . as would any other scientist who can appreciate the pain and embarrassment of having a great finding, a breakthrough, proved false because of a technical experimental error. I am happy to report that Mrs. Adie carried on. Two years later we find

15. Try as I might, I have been unable to trace the person of Mrs. Adie. She was a medical protozoologist, as were several other distinguished women scientists in the era between the two world wars (and also today). She had, however, the singular distinction (as far as I know) of being the only woman scientist of that time actually to work in the tropics on tropical diseases. She must have been a wonderful character and I wish I had known her.

her publishing an article on a malaria-like parasite of pigeons, a subject that could not excite great controversy or strong passions except amongst a small coterie of purists.

As the bedbug theory was being discredited, others were casting about for new candidates. One of the men who was to put his formidable energies to the effort was Major John Sinton of the Central Research Institute's Medical Entomology Section at Kasauli. Sinton, a North Irishman, was to become renowned not only for his stature as a malariologist but also because he would be the only person to hold both of Britain's highest honors, the Victoria Cross for military galantry and Fellowship of the Royal Society for scientific achievement.[16] At that time, it was not beneath the dignity of physicians to study insects, and most of the foremost medical entomologists were doctors who combined an expertise of medicine and zoology. In considering the kala azar vector problem, Sinton did what any good military man would do: he looked at maps. The map of kala azar showed its restricted range in eastern India, from Madras to Assam. When distribution maps of the blood-sucking biting insects were overlayed on the kala azar map, the range of one species neatly coincided—*Phlebotomus argentipes,* the silvery sandfly. In 1924 and 1925, Sinton published papers advancing his theory that the sandfly was the vector of the kala azar parasite, *Leishmania donovani.* The pursuit of the sandfly was now joined. It would take another twenty years

16. During a somewhat careless life in the tropics and elsewhere, I've had several outstandingly frightening experiences—for example, with a hippo who thought I didn't belong in his water, and the confrontational terrors of the African road. But perhaps the most unnerving experience was, as a very young graduate student, to be the bridge partner of Brigadier John A. Sinton, V.C., F.R.S. "The Brig," a wonderfully kind man, would bid with the same panache that won him his V.C. Mostly three no trump, as I recall. And heaven help his partner who didn't make the contract, no matter what cards he held.

before the final piece of conclusive proof was put into place. But at least the trail was now true.

The Calcutta School of Tropical Medicine had an additional piece of epidemiological evidence incriminating the sandfly—Ward 14. Ward 14 of Calcutta was largely populated by another product of British colonial rule, the Anglo-Indians. Merle Oberon and Ava Gardner's Bhowani Junction apart, the Anglos led unromantic lives in an ambivalent subculture that was neither quite English nor quite Indian. In expiation for their sins of miscegenation, the parental English virtually bequeathed the Indian railway system upon their mixed-race progeny, and it was run by them with great efficiency. The Anglo-Indians, in making a somewhat distorted cultural alignment, built homes which they thought of as being typical "mother country"—large, morose-looking wooden houses within a compound of thickly shadowing foliage. In 1925, the Anglo-Indians of Ward 14 were dying of kala azar. Their distant Hindu cousins of Calcutta's northern wards were barely affected by the disease. Searching for the reason for this difference, the Calcutta School of Tropical Medicine scientists headed by Robert Knowles noted that the dark rooms of the Anglo-Indian houses, "lumbered with family furniture," humid from the surrounding dank vegetation, were optimum conditions for the livelihood of *Phlebotomus argentipes.* The Anglo-Indian houses held large populations of this sandfly, while the simpler, more open shacks and homes of the Indians did not. In the poorer Indian wards there were lots of fleas, lice, and bedbugs, but relatively few sandflies. This finding encouraged the Calcutta School group to begin the demanding transmission experiments.

None of us likes sandflies. They are extremely finicky to breed and maintain in the laboratory. Identification of their species often requires such expert practice as dissecting out their genitalia—and they are very small creatures with very small genitalia. Despite these difficulties, which were even more rigorous in 1925 than now, the Calcutta scientists managed to establish a thriving colony of *Phlebotomus argentipes*. Knowles began to feed these "clean," laboratory-bred sandflies on kala azar patients; then, day by day, he took the flies apart. On a glass slide under the microscope each fly was carefully dissected and searched for the presence of the flagellate form of *Leishmania donovani*. There was a considerable sense of excitement when these forms were seen in the gut of flies that had been fed on the patients three or four days earlier. Twelve days later the flagellate forms had multiplied and were now in the "throat" of the sandfly. This was great progress, but it still didn't prove actual transmission. The crucial A to B experiment had yet to be performed: an infected sandfly had to bite a human "guinea pig," and that human had to come down with kala azar.

The protracted transmission studies were now largely undertaken by officer-scientists deputed to the Indian Kala Azar Commission. The original trio, who set up shop in Assam, were Colonel Rickard (later Sir Rickard) Christophers, Major Henry Edward Shortt (later Professor Shortt), and a Mr. P. J. Barraud whom we don't hear much of as a "later." Knowles communicated his findings to the Commission, and Christophers, et al., quickly confirmed the Calcutta group's findings. Looking into their overly optimistic clouded crystal ball, the Commission predicted in their First Report of 1926 that "Only experimental transmission by the sandfly would therefore now seem to be necessary to prove finally the *role* of this insect in the transmission of kala azar." Little did they then know that the successful A to B experiment would take another fourteen years.

Henry Edward Shortt, the only one of the trio to stay the course, would never carp over what must have been frustrating years. He was a man who pursued, and pursued, his quarry. He loved the hunt. As an Indian Medical Service officer, he shot tiger. As a professor at the University of London's prestigious School of Tropical Medicine and Hygiene (where he was to discover the missing "liver" link of the malaria parasite's life cycle), he taught his graduate students, at teatime, how to stalk and kill houseflies by creeping up on them with two extended fingers. As a retired visiting professor emeritus in Africa, he hunted agamid lizards (to study their malaria) with a blowpipe he made from glass laboratory tubing and modeling clay as the pellet missiles. No Jivaro could have been more accurate than this deceptively mild-looking man of modest stature peering at his prey over half-glasses. At the age of one hundred he caught his last trout and died, leaving his widow, his beloved Hobby, to survive him for another year until she joined him in death at age one hundred and three (Knowles had been the best man at their wedding in India those many years ago). Shortt knew that it was now no trick to produce infected flies; but try as he and his colleagues did year after year, time after time, the bite of those flies did not produce infection in human volunteers. Some experimental technique, some trick, was missing. When the trick was *finally* discovered, it seemed so simple; it also showed how scientists could be snookered for so long behind entrenched, plausible—but erroneous—assumptions.

The assumption was that the sandfly was rather like a small mosquito in its dietary habits. The male mosquito and the male sandfly, gentle souls, are vegetarians, feeding exclusively on fruit juices and other plant material. They partake of no blood. Only the ladies are blood feeders. This is certainly true of mosquitoes but, as it turned out, not quite true of the sandfly. In 1939 a physician-cum-entomologist, Dr. R. O. Smith, established a laboratory in Bihar to study sandflies, beginning the research that was so important in unraveling the transmission knot. First, Smith showed that the female sandfly would readily feed on fruit (Smith gave them raisins) after taking the initial, infecting blood meal.[17] Next, he showed that when the usual procedure was carried out of maintaining the flies on "clean" blood feeds (as in malaria transmission techniques with mosquitoes), the second blood feed would, for some inexplicable reason, halt the multiplication of the flagellate *Leishmania* parasites. It caused them to lose their vitality, and sometimes even wiped them out from the sandfly.

Even more startling was Smith's observation that the raisin diet following the infective blood meal caused the flagellates to thrive and multiply so enormously that the sheer numbers of organisms formed a plug in the sandfly's pharynx. It was these "plugged" or "blocked" sandflies that constituted the potential transmissive danger. When they tried to feed again, the plug of parasites occluded the throat and made feeding impossible. The sandflies made violent efforts to ingest blood or other fluids, and in doing so, some of the flagellates of the plug became dislodged. It was these

17. My wife says that one of her most prominent memories of a sabbatical year I spent at the London School of Hygiene and Tropical Medicine was the scene at the morning coffee break, the "elevenses," where from time to time she joined me and the other faculty members. The entomologists, she vividly recalls, would turn up in the coffee room with cages of mosquitoes strapped to their arms and legs. They would nonchalantly have their "elevenses" of coffee while the female mosquito companions had theirs of blood.

dislodged organisms that might be capable of infecting a human when the "blocked" sandfly tried to feed. Shortt and all the other researchers had been following the malaria-mosquito experimental procedures of successive blood feeds. With the sandfly *Leishmania donovani*, this gave rise only to non-infective parasites.

In 1940 an Indian physician-scientist, C. S. Swaminath, and Shortt made the successful A to B experiments.[18] Swaminath collected six Indian volunteers from a hill district of Assam and fed the "raisined" infected flies on them. Three of the volunteers contracted the disease. The sandfly, *Phlebotomus argentipes*, was finally confirmed as the vector of kala azar. Thirty-eight years to prove a point! Only the post-game criticisms customary for new discoveries had still to be dealt with.

There was one bit of sniping that was amusing and would be impossible in modern science's impersonal, peer-reviewed publication policies. In 1944, a Dr. Malone published a letter in the *Indian Medical Gazette* in which he advanced his doubts that Shortt and Swaminath had proved their case for the sandfly transmission of kala azar. A few weeks later Shortt makes his Shorttian reply in the *Gazette:* He knows Dr. Malone to be a disciple of George Bernard Shaw, therefore the opinions of a fervent Shavian socialist cannot be accepted. Finis. At a much later year, Shortt was to refer to a well-known but not-quite-first-class scientist as "The Bishop" because he was always confirming others (including Shortt's recent discovery of the liver phase of the

18. From the 1920s onward it was not all a "one-man show" of British scientists; many competent and distinguished Indian physicians were engaged in kala azar research and often took precedence in publication. Shortt always spoke highly of his former Indian colleagues, although he might occasionally grumble that in the midst of a crucial experiment they would depart on a holy pilgrimage for an indeterminate period.

malaria parasite). Knowles, in his lecture notes from the Calcutta School, sums up the protracted search with a kind of eulogy of exhaustion:

> The story of the discovery of how kala azar is transmitted from man to man is one of the most amusing, also perhaps one of the sorriest in tropical medicine. It is a history of almost twenty years of wasted effort, of individual workers starting off with the highest hopes and ending in despair; of false starts and erroneous conclusions; of acute controversies and the flow of much ink; of wasted effort and the absence of co-ordinated enquiry.

Chapter 5

New Knowledge, New Treatment —and New Epidemics

NOW IT WAS known how you got kala azar—from the bite of an infected sandfly. Knowledge alone, however, couldn't prevent the insect's bite, and the epidemics of kala azar continued. Between 1918 and 1923, more than 200,000 people died of the disease in Assam and the Brahmaputra River Valley. In 1944, another epidemic struck the region. The sick and dying with their gray pallor, great spleens, and emaciated bodies continued to come to the hospitals and clinics. But from 1913 onward there was some hope. There was a curative, albeit imperfect drug: a slow and toxic not-so-magic bullet that derived from the vanity surrounding an ancient woman's eyes.

Egyptian women of the early dynasties and, probably, their sisters of several hundreds or thousands of years before, enhanced their beauty with a dark blue mineral that they crushed into a powdery paste. These ancient cosmetics were oxides and sulfides of the heavy metal antimony. One of these ladies, a long, long time ago, must have noticed that her cosmetic healed a sore on her face. She may not have been a very "nice" lady and it may have been a syphilitic sore, a skin disorder against which antimony oxide has a therapeutic effect.[19] Heavy-metal pharmacology thus began in ancient times and enjoyed a vogue as a panacea until the Middle Ages. Elegant goblets were made of antimony and wine drunk from them as a curative practice in the way we now take our daily vitamin pills. Antimony seems to have fallen from medical favor in the 1400s but had a brief revival when a quack successfully treated King Louis XIV with an antimony compound for some unknown illness. Then, in the first years of the twentieth century, the German pharmacologists, led by the genius of Paul Ehrlich, began to formulate arsenical and antimonial compounds that gave reasonably good curative effect against syphilis and African sleeping sickness. Attention again turned to the possibilities of heavy-metal therapy.

In 1903, Sir Leonard Rogers began infusing antimony oxide into kala azar patients. It required a two-month course of this toxic therapy to get a few cures. In 1915, a new antimony compound, antimony tartrate (tartar emetic), was introduced. It is a vicious drug. One practitioner described it as ". . . not a pleasant drug; it produces cough, chest pain, and great depression just after injection, so that patients have often refused to continue with it . . . it is a definite poison with action on the heart, for one case died of sudden heart failure within an hour of injection, and others appeared to be hastened towards heart failure by it." And it wasn't much good in the treatment of kala azar. Shortt made his characteristic mordant observation in declaring that it might be better than no treatment at all, but 90 percent of the treated cases died.

19. Actually, it probably was an infection other than syphilis. Most medical historians are of the opinion that syphilis had its origins in the New World and that at the time of the Egyptian dynasties the infection was still confined to the American continents.

All these compounds were made of antimony in a molecular state known as trivalent. Trivalent antimonials, as I have noted, are poisonous. Sometimes they kill the parasite. Sometimes they kill the patient. Also, the body doesn't excrete these compounds very rapidly and so the toxicity tends to build up and be prolonged. But it is remarkable what a difference a few electrons can make. In the 1920s compounds of antimony in another molecular state, pentavalent, began to be synthesized and introduced for the treatment of kala azar. Finally, in 1935, a pentavalent antimonial known as Pentostam was produced. Despite all its drawbacks this was the most effective drug to that date. Any date, for that matter. Even after more than fifty years it remains the essential chemotherapeutic for kala azar. It is the drug that Susheela sought for her dying child.

While it is a wonderful advantage to have the therapeutic means to cure the sick, it is an even more wonderful advantage to have the means to prevent them from getting sick in the first place. Pentostam may have cured the patient, but it did nothing to affect the breeding and biting sandflies. In 1935 there were no powerful, long-acting, inexpensive insecticides. "Quick, Sanjay, the Flit" would kill on contact, but an hour later another spritz would be necessary. At any rate, the Sanjays of the poor tropical world couldn't afford the Flitlike insecticides of that period.

It may well be that the two products of greatest importance to emerge from the Swiss Patent Office were Albert Einstein and patent #226,180. Albert Einstein excited our intellect and made us view our universe in an entirely new way. But it was #226,180 that saved millions of lives. #226,180 was DDT. In 1940 Paul Muller, a Swiss citizen, registered his discovery of the first chlorinated hydrocarbon insecticide that we know by its acronym of DDT. There was nothing quite like it before and has been nothing quite like it since. Here was a chemical that could be sprayed on the walls of a house and for up to six months later any insect that alighted or rested on that wall would die.[20] It was virtually without toxicity to humans. And, for the icing on the chemical cake, it was dirt-cheap to manufacture. For the first time, medical science had a weapon which allowed the vision and hope that insect-transmitted diseases could not only be controlled but *completely eradicated.*

In 1940, malaria was the leading cause of sickness and death of humans; and not only in the steamy tropical lands. It was a great health problem in Europe, north to Holland. In 1940, malaria was, in many ways, *the* American disease. It is significant that one of the first major health applications of DDT was the large-scale trial in 1945 to control malaria along the Tennessee River. The success of that trial launched the Global Eradication of Malaria Plan that emanated from the World Health Organization's headquarters in Geneva a few years later. The promise of DDT was so great that in 1948 the Nobel committee awarded its Prize to Paul Muller on behalf of the anticipating and grateful human community.[21]

I will have more to say about the malaria global eradi-

20. DDT doesn't act as a "knock-down" as pyrethrum does. It is picked up by the insect through its outer chitinous "skin" and then acts on the insect's nervous system. It may take several hours, or longer, before the insect will die of the "DDT jitters."

21. Muller may have got "the Prize," but actually he was not the discoverer of dichlorodiphenyl trichloroethylene (DDT). A little-known and usually overlooked fact of that matter is that DDT was first synthesized in 1874 by a Viennese pharmacist with the name of Othmar Ziedler. During World War II, Muller was employed by the Swiss chemical-pharmaceutical firm Geigy, in the neutralist pursuit of trying to find a chemical to control clothes moths. He resurrected Ziedler's compound and found it to have just the activity he was seeking. He then passed some of it on to the U.S. Army.

cation program and its ultimate failure in a later chapter; in this chapter I will merely outline the basic strategy of the program. This was to spray *all* human and animal habitations with DDT twice a year for at least five years. It was calculated that the mosquito population would thereby be reduced to a level where malaria transmission would no longer occur. India bought the scheme and established its National Malaria Eradication Program in 1952. India supplied the willingness, determination, and its national health resources. Its then good friend Uncle Sam bankrolled the project and supplied the DDT, which was being manufactured in the United States at a rate of over 200,000 tons a year (at its peak usage America produced over 400,000 tons a year). And everybody's good friend, the United Nation's World Health Organization, provided advice and expertise.

However, the great unexpected additional bounty of the antimalaria spray program was its effect on kala azar. Considering the natural history of *Phlebotomus argentipes* it is only natural that DDT would be as effective, or more so, against the sandfly as against the mosquito. *Phlebotomus argentipes* is a domestic creature. In India, the cow is also a domestic creature and it is the sandfly's life support system. Sacred cows have mundane manure. Susheela's cow byre, you may recall, was attached to her house. It was humid, dark, and carpeted with a rich layer of dung. These are the ideal conditions for *Phlebotomus argentipes*. The female deposits her eggs on the manured floor; the larvae hatch and feed on the organic manure matter. Some days later, the satiated larva cocoons into a pupa, and a week or so after that the adult sandfly emerges. The sandfly is not a strong flyer; and why should it be? The necessities of life—

food and sex—are all to be had at its birthplace. There were lots of flies of the opposite sex for easy nuptials and a constant source of blood food from both the resident cow and the Devi family. The sandflies led their quiet lives of feeding and fornication. They were fruitful and they multiplied. And they rested on the walls inside the house, usually not above six feet from the floor.

These were habits that made them ideal targets for DDT. When the spraymen came to kill the malaria-transmitting anopheline mosquitoes, the kala azar-transmitting phlebotomine sandflies got a dose of DDT, too. The DDT was especially effective against them because they tended to rest on the wall more often for longer periods and at more accessible heights than the anophelines. They were also (and still are) exquisitely sensitive to DDT, being killed by concentrations that would be too low to affect the mosquito. In some places a single spray round would suppress the *Phlebotomus argentipes* population for an entire year. The malaria program devastated the sandfly population. Transmission of kala azar was interrupted. By the mid-1950s there were virtually no new cases of kala azar. By 1965 it was a forgotten disease. In 1970 kala azar returned—it was 1940 all over again.

During the first years of the 1960s it became all too apparent that malaria could not be eradicated by the national campaign, and in 1970 or 1971 the Indian government decided to disband the malaria eradication program. That decision was probably also influenced by the introduction of a cheap, highly effective antimalarial drug, chloroquine. It was becoming more and more expensive and frustratingly ineffective to control the mosquito. Handing out the drug was much cheaper. Chloroquine didn't eradicate malaria, but it did largely stop the deaths and, for the most part, it prevented people from getting too ill.[22] Without the containment of DDT the sandfly population proliferated to its former great numbers. In 1969 or 1970—no one is certain of the exact date—the inevitable came to pass. In the village of Vaishali, people began dying of kala azar.

Vaishali is a modest and typical agricultural village of Bihar State. It is a village with a pleasant, muted quality of peace and serenity. There is shade from the many mango trees clothed in their British racing green leaves. There is a "tank"—a large rectangular man-made pond filled with algae-thick water. A government rest house sits by the side of the tank, a temporary refuge for the touring officers of the colonial past and now, infrequently, for officers of the Indian Civil Service. There is a great Bo tree by the rest house and a garden which in the late fall is bright with the asters, marigolds, roses, and dahlias from earlier plantings. In the late morning, when the sun has warmed them, a flurry of delius butterflies brightens the flowers; their flashy underwings of orange and black making a cloud of moving color. And in the morning you can sit in an old wooky chair on the rest-house veranda, have tea, and satisfy the inner need with peppery samosas bought, still hot, from the itinerant vendor. Not a bad place to be, Vaishali. And if ever there was a village that should have enjoyed the protection of divine intervention this was it, for in Vaishali the Buddha is reputed to have experienced his last enlightenment before his death. It is the birthplace of a reversed Jain saint. But

22. As I'll note in chapter 14, chloroquine wasn't the final solution, either. By the late 1960s and early 70s more and more strains of the deadly malaria parasite, *Plasmodium falciparum*, were becoming solidly resistant to chloroquine. Nor was there another antimalarial drug of equal efficacy and price to replace it.

it was in Vaishali that kala azar returned to begin claim to its ancient domain.

In some respects Vaishali was like Jessore, that other epicenter of kala azar some 150 years earlier. When kala azar struck Jessore in 1724, it seemed to be a new visitation to the people of that city. By 1972 the villagers of Vaishali had lost their memory of kala azar—twenty years had passed without a single case—and when it struck, they did not recognize its insidious signs: the fever, lassitude, and urgent frequency of diarrhea. These are common complaints in rural India; only by their persistence do they gain the distinction of being the early announcements of kala azar. The stricken were at first perplexed when they did not recover as they usually did; the elixirs, pills, and tins of ointments obtained from traditional or conventional doctors or "over-the-stall" in the village market were of no help. It was, perhaps, a year or more before the people of Vaishali realized that they were besieged by a new epidemic.

In 1724, the medical men of Jessore had no knowledge of kala azar or how to treat it. In 1972, the doctors of Vaishali had all but lost *their* knowledge of kala azar. Their medical training had given scant attention to this oddment of parasitology. To see a case during their clinical years would cause the same curiosity and excitement as one of our American fourth-year medical students being shown a case of diphtheria. So, when the sick citizens of Vaishali came to their doctors with the early symptoms of kala azar, they were given a pill for malaria—the tropical counterpart of "take two aspirin and call me in the morning." Then, when the Vaishali patient returned a week later still feverish, he or she was given the customary government-issue antibiotic, tetracycline or penicillin. It was the increasing number of persistent patients with their persistent fevers that finally made the physicians realize something unusual was occurring. A year or more was to pass before the doctors of Vaishali became aware that they were in the midst of a new epidemic of kala azar.

In 1724, there was no drug to treat kala azar. In 1972, the doctors again had no drug; the pharmaceutical companies had all but stopped the manufacture of the antimonials. There is no profit in making drugs for diseases that barely exist. For those dying of kala azar in Vaishali in 1972 it might have seemed that nothing had changed in 250 years.

Kala azar was not to be contained. By 1977 it had spread from Vaishali and nearby Muzaffarpur throughout most of Bihar State. In 1975, the infection made its way to adjoining and densely populated West Bengal. By the early 1980s there were disturbing reports of outliers of the disease in Uttar Pradesh and the Dravidian homeland of Tamil Nadu.

National boundaries are constructs of human political imagination; pathogens need no visa for their journeys. Kala azar made its way across the river and into the plain of Bangladesh sometime during 1979 or 1980. People, mostly children, began to die in Pabna, a district not more than 100 miles from the national capital, Dhaka.

We imagine—and we are told—that all Bangladeshis lead lives of unrelieved, desperate squalor. This is not so. There is entrenched grinding poverty, particularly in the cities overcrowded by emigrants dispossessed from their rural villages by overpopulation. Yet the villagers who remain do manage. They are not prosperous, but they do have the renewing organic richness from the Gangetic rivers, and even the cataclysms of tornado and flood bring new farming land in their wake. The national wars are behind them and the country people survive—tough, bright, good-humored

Bengalis. They even have reasonably good access to medical services. Actually, only the access is reasonable; it's the service that leaves much to be desired. In the 1970s the Bangladesh government created many new medical schools which produced many new doctors who did not have many new jobs to go to or a paying patient base to support them. Rather than have young doctors revolting, the government (already politically beset by parties of many stripes) decided to employ them all and dispersed them to health centers throughout the country. Then the government found that it had spent virtually all of its health budget on doctors' salaries (which wasn't all *that* much; it was just that there were a lot of doctors and not a very big budget) and there was little left over for the purchase of medicines and supplies. When the people of Pabna brought their stricken children to the local health center, the doctors were usually able to make an accurate diagnosis of kala azar. Diagnose was all they could do; they had no supply of the antimonial drug. During the early stages of the epidemic there wasn't even any drug on the market. So parents took their children home to await their death. Eventually, they recognized the futility of it all and stopped coming to the health center. In the midst of the kala azar epidemic the doctors of Pabna waited for patients who never came.

Finally, kala azar crossed its last national boundary to enter the Terai, the agricultural plain of Nepal bordering India's Bihar State. There were few health centers in the Terai and no drugs. The sick with money made their way to Bihar or Calcutta to be treated in private clinics. The sick without money are largely a lost statistic.

Gathering the lost statistic has caused great difficulty to those trying to deal with kala azar. Even in industrialized countries, the epidemiologists—for all their computers, reliable laboratory support, mobility, laws-of-the-land requiring that notifiable diseases be reported to a government health agency, and full-time jobs on a livable wage—have their problems in acquiring accurate data on health and disease. The information gathered by epidemiologists provides the basis for the logical planning of antidisease campaigns.

Third World epidemiologists rarely have these supports. Nor are they as respected as "real," hands-on-the-patient doctors. All too often, they are physicians who were unable to secure advanced training in a clinical specialty and opted for a career in public health epidemiology as the only opportunity to escape the general physician pool. With any luck, they would get the ultimate escape of study at an overseas school of public health. All too often, their functions are diluted by administrative or teaching duties. And all too often, economic circumstances limit them to being "armchair" epidemiologists. Like all government physicians, their pay is inadequate and almost all have private practices in the afternoons and evenings. An epidemiologist, to be effective, must go to the disease outbreak to grub and probe for information for as long as necessary. The Third World epidemiologist, bound to his private practice, is usually limited to day trips away from his city office. The few who are uncommitted to private practice still experience great difficulty in undertaking field studies. There is a shortage of vehicles, a shortage of gasoline, a shortage of travel money, and a shortage of sympathetic administrators supportive of the notion that epidemiology is not a sedentary profession.

Of course there were "numbers." Governments always have numbers. When you thread your way through the labyrinthine West Bengal Health Department Building in

Calcutta—through the upstairs and the downstairs, switchbacks and roundabouts, outside verandas and inside balconies, past armies of clerks and mountains of files bound with puce-colored cloth tape—you may eventually find the state epidemiologist. He is a pleasant man who has been around the track more than once. He knows the game. His walls are covered with charts and tables of numbers, graphs of incidences and prevalence rates. The kala azar chart for last year shows 4500 cases and 125 deaths in all of West Bengal; a minor infection compared to malaria or the diarrheas. In his more candid reports he gives the estimate as 14,000 to 20,000 new cases that year, with 1500 to 2000 deaths for all of West Bengal in that year. Not a minor infection by *those* statistics.

Five years before, when the number of cases in West Bengal appeared to have increased precipitously, the state government was stirred into carrying out a DDT spray round of the worst-hit villages. This proved to be an exceedingly difficult operation. First of all, the DDT sent to the state epidemiologist wasn't very potent. It had been manufactured, he believed, in Pakistan and had sat around for a long period on the dock, exposed to sun and rain, before being shipped. And according to the Health Department's chemist it wasn't very pure, even before it had undergone the deteriorative conditions of storage—"pure rubbish," as the chemist characterized it. As an aside, the epidemiologist said he hoped that the Pakistani atomic bomb (he was convinced they had one) would be no more potent than the DDT they manufactured.

If the DDT was rubbish, so was the spray equipment abandoned by the National Malaria Eradication Program and left, uncared for, to deteriorate. Even after extensive cannibalization there still wasn't enough equipment in good working order to do the job properly. Perhaps the most serious problem was the lack of spraymen willing to work. Under the National Malaria Eradication Program the spraymen had been well-trained, well-supervised, salaried employees with assured "federal" jobs—pukka Government Men. That status and those jobs all came to an end when the National Malaria Eradication Program crashed. Now when they were offered temporary jobs by the state government they again wanted to be permanent employees and most disdained the offer to work for an hourly wage. As a consequence, that one spray round was the last and only effort to control the sandfly vector and contain transmission. It was a feeble effort; but compared to Bangladesh and Nepal, India's was a model of epidemiological efficiency.

Bangladesh did its kala azar epidemiology by official denial. As early as 1978 or 1979 reports of an outbreak of kala azar filtered through to the Directorate of Health in Dhaka. One could almost hear the official sigh of desperation; was it not enough that they were trying to keep up with cholera and every other known intestinal disease as well as respiratory infections of many varieties, malnutrition, population control,[23] and sundry other illnesses? Small wonder that they turned a deaf ear to the reports of kala azar in the village 100 miles from the capital and a long day's journey across the Brahmaputra. Almost by default, kala azar became the responsibility of a semigovernment academic institution, the National Institute for Preventive and Social Medicine, better known as NIPSOM.

23. The population of Bangladesh, now about 110 million, is expected to double within twenty-seven years. Meanwhile the infant mortality rate is, by the official statistic, 135 per 1000, and for those that survive, the life expectancy is fifty years.

NIPSOM is somewhat like American schools of public health, but in addition to an extensive teaching program NIPSOM is responsible for research and surveillance of diseases of public health importance in Bangladesh. When the Institute began work on kala azar, the director, Dr. A. K. M. Kafiluddin, was an eccentric who would break off conversations to call his psychiatrist (possibly the only shrink in Bangladesh) to berate him for not preventing the anxieties and depressions of that moment. Then, with cool aplomb, he would refocus on his visitor and continue the discussion. Later, he built a new research wing at NIPSOM, consulting no one, with laboratories that had no water or gas and an electric power supply sufficient to operate little more than a toaster.

In the powerless, waterless laboratories there was a feast and famine of resources. Donors were lined up, six deep, cash in hand, filling the hotels of Dhaka, to fund health programs. The funds that NIPSOM received allowed the departments favored by the director to order in "one-of-everything" style from the supply house catalogues. Usually, the equipment and supplies—all perishable within the fullness of time—would languish on the docks of the port, Chittagong, and then eventually disappear for months, years, into the caverns of the government storehouses. For some unfathomable reason, the Chittagong Customs officers were convinced that the instruction manuals were subversive literature. The crates and packages were opened by Customs and the manuals, the "books of words," were confiscated. Equipment and supplies for laboratory and epidemiological studies, if and when they did turn up at NIPSOM, frequently could not be assembled or used for lack of instruction manuals.

For all practical purposes, Bangladesh is run by the military, and when everything else is in shambles the military usually run their own establishment with efficiency. A World Health Organization short-term consultant assigned to NIPSOM who had spent a day of crazy-making frustration trying to assemble a $7,000 analytical balance without benefit of the manufacturer's how-to manual complained, at a dinner party that evening, to the colonel in charge of medical stores for the Bangladesh army. The colonel was very sympathetic. And for good reason; the Customs of Chittagong took the manuals from his equipment also. Those guardians were not going to let such inflammatory literature subvert their nation's armed forces.

So, with a loony director, laboratories without water and power, one-of-a-kind supplies (if a microscope bulb burned out, it was necessary to get a new microscope from the stores—there were no replacement bulbs), and apparatus that was often improperly set up or used, NIPSOM's epidemiologists would have had considerable difficulty in getting the needed laboratory support for any kala azar studies they might undertake. It really didn't make any difference because the NIPSOM epidemiologists weren't going anywhere. First of all, their teaching obligations were so extensive that they had little time (after their early departure to private practices was subtracted) for work in the field. Then, the director became agitated when any of his staff were beyond peremptory call, and he discouraged field investigations. They wouldn't have gone, anyway. As one young epidemiologist explained, this was a Moslem society; his wife was in purdah—not completely confined to the house but she certainly couldn't venture out without her husband. She couldn't do the marketing. That was the epidemiologist's domestic duty. His family would starve if he had to leave Dhaka for more than a day or two to carry out surveillance work in the countryside.

Remarkably, within this morass there surfaced a NIPSOM faculty member who was determined to undertake epidemiological studies of kala azar in Bangladesh. Dr. Nurul Islam Khan, who had experienced his share of Bangladeshi vicissitudes, was determined to become a parasitologist. As a young, recent medical graduate he had to flee Dhaka with his new wife and their infant child during the war of independence from Pakistan. The Khans' flight took them on a small boat drifting down the Brahmaputra. On this journey they were disguised as farmers because the Pakistanis were bent upon exterminating doctors, teachers, and others of the educated class as part of their policy to retain control of what was then East Pakistan. After the war ended and Bangladesh had become an independent nation, Khan was sent to the London School of Hygiene and Tropical Medicine where he was given basic training in parasitology. When he returned, he was posted to NIPSOM and became head of the Department of Parasitology. His determination and devotion to science can be measured by his being, alone amongst the NIPSOM staff, the only really full-time worker. He somehow managed to live on his government salary unaugmented by private practice. Nurul Islam was aware of kala azar's notorious past in Bangladesh (East Bengal which became East Pakistan which became Bangladesh) and he was also aware of what was now happening in the bordering Indian states. If no one else would, he would investigate the reports and rumors of a new outbreak of kala azar in Bangladesh.

Nurul Islam did not lack determination, but he did lack the wheels to do field work. The NIPSOM garage was littered with old, semi-old, semi-new, and almost-new vehicles, all inoperable or so undependable as to be unfit for travel beyond the capital. The first budget item for foreign-financed health projects was, usually, vehicles. Like the microscopes, spares for the vehicles were not ordered. Fortunately for Nurul Islam, the regional office of the World Health Organization in Delhi[24] had assigned a Czech epidemiologist on long-term contract to NIPSOM. This man was a study in frustration. None of the Institute epidemiologists would do anything other than teach; no one followed his sage WHO-anointed consultant advice. Dr. Nurul Islam Khan was as a gift from Allah. The Czech didn't know much about kala azar but he did know a live one when he saw one. Here was a senior NIPSOM staff member who actually wanted to study a disease that might be of importance in Bangladesh. Nurul Islam got the favored-son treatment. Arrangements were made for him and his team to use World Health Organization vehicles (there is a WHO office and consultant group in Bangladesh); urgently needed spares and reagents came through the World Health Organization's diplomatic pouch. When technical expertise was required, Nurul Islam got the short-term consultant of his choice.

Dr. Khan did not fail the faith entrusted to him. His studies in the Pabna district rapidly confirmed earlier reports and rumors of the presence of ongoing kala azar transmission. Within a few weeks of beginning work he had detected several hundred new cases. In small, primitively equipped health centers he performed diagnostic bone-marrow biopsies and was able to bring cultures of *Leishmania donovani*,

24. The World Health Organization has a central headquarters and a grand edifice in Geneva. There they hold countless meetings of countless committees and issue a tonnage of reports. They try to promote research and carry out a few projects in the field. The real power of the World Health Organization has shifted to the regional offices, semiautonomous bodies whose heads are elected within the countries of the region. The day-to-day action, trying to deal with local health problems, goes on at the regional level. But as you might imagine, there is lots of politicking and payback.

seeded from infected bone marrows, back to his laboratory in Dhaka. he began to develop a competence in serological diagnosis—looking for the specific antibodies in the blood which would allow him to undertake extensive epidemiological studies. Khan was appalled by the almost total absence of anti-kala azar drugs, and with dogged persistence he went after the government and drug company representatives to get the needed antimony gluconate and then to get it to where it was needed. Then Nurul Islam Khan died.

There is a tragic thread touching the personal lives of several medical scientists investigating kala azar in the Indian subcontinent. When Nurul Islam's father-in-law died, Nurul Islam and his twelve-year-old son, faithful to custom, went to the banks of the Brahmaputra to wash the corpse in preparation for burial. It was monsoon season; the river was in violent spate. Nurul Islam's son ventured a little too close to the water's edge and was swept away by the swift current. Nurul Islam couldn't swim a stroke but without hesitation he went to rescue his son. They both drowned.

The bright promise of practical inquiry into kala azar in Bangladesh died with Nurul Islam Khan. The Czech epidemiologist's contract was not renewed, nor was the epidemiology program which supported the long-term assistance to NIPSOM. The Institute replaced Nurul Islam with another medical parasitologist who also wanted to "do something," but he was new to kala azar studies and the committed technical support he needed was not forthcoming. The policy of the World Health Organization's regional office in Delhi subtly changed. The expert consultants, formerly selected from an international panel of distinguished and experienced biomedical scientists, were now largely chosen from within the region. It was cheaper and certainly more politically expedient in providing "jobs for the boys," but it didn't help the people like our "green" parasitologist from NIPSOM. To his credit, he did get into the field once or twice and he did find that kala azar had spread to districts from which it had not been previously reported. Then he too joined the limbo of unrequited researchers of South Asia.

There isn't a lot one can say about Nepal, the third country of the Indian subcontinent in which kala azar had again become resurgent. Nepal's majestic geography belies a poverty that equals, if not exceeds, that of Bangladesh. Medical services exist but remain rudimentary. Nepal's first medical school came into being only a few years ago. There were few medical specialists, fewer epidemiologists, and fewer still epidemiologists willing to travel beyond the capital, Kathmandu.

The immediate associative image of Nepal is of immense mountains, the top of the world. That is, of course, true; but there is another Nepal. To the southwest of Kathmandu a treacherous Chinese-built road descends from Mogul's Bazaar to the Terai, the Gangetic plain of Nepal. The Chitawan Game Reserve, Nepal's last remaining enclave of forest and wildlife, is situated where the mountains meet the plain. After Chitawan, there are enormous fields of gold—flowering mustard plants, stretching to the horizon. From time immemorial, the Terai has been notorious for the constant intensity of malaria and intermittent resurgence of kala azar outbreaks.[25]

Beginning in 1980, the health centers in the outlying districts of the Terai began seeing cases of kala azar. This

25. Development of *Leishmania donovani* in the sandfly requires an optimum warm temperature. Although sandflies exist at higher elevations, it is too cold from about an altitude of 1500 meters for the parasite's developmental cycle to be completed in the insect.

was new to their experience and they reported the occurrence of the disease to the Department of Health's Division of Epidemiology and Statistics in Kathmandu. That division, in turn, appointed a veterinarian of the Zoonotic Disease Control section to monitor the kala azar situation. Why a veterinarian to monitor a purely human disease? Evidently the Nepalese clung to the mistaken belief that the Indian form of kala azar was like that of the disease in other endemic parts of the world, a zoonosis with wild and domestic canines as reservoirs. More than fifty years of negative results, in which countless numbers of dogs, jackals, and all sorts of other animals had been killed and their tissues searched for evidence of the parasite, had not completely discredited the hypothesis. It was still thought that veterinarians should play an important role in control of the Indian *Leishmania* parasite.

From the health centers the government learned that in 1984 there were 605 cases of kala azar in the Terai. This was meaningless as an epidemiological statistic in the way that data from India and Bangladesh had been meaningless. Nor was there any stock of antimony gluconate in Nepal. Even fewer people came to the government health centers—the source of the reported case numbers—for treatment. Beyond dispute, however, was that kala azar had returned to the Terai.

Chapter 6

The World Health Organization to the Rescue

BY 1985 there was a kind of semi-official recognition that (1) kala azar was now endemic and spreading in the three countries of the Indian subcontinent—India, Bangladesh, and Nepal; (2) there were insufficient stocks of antimony gluconate for widespread therapeutic application;[26] (3) little or nothing was being done in the way of sandfly vector control; and (4) insufficient funds and trained medical personnel existed to undertake an anti-kala azar campaign even if the official hearts were in the right place. To this parlous state of affairs came Dr. Sadnand Patnayak. Patnayak had been the head of India's malaria eradication program and when he finally admitted to defeat by the malaria parasite, the anopheline mosquito, and obstinate officialdom, he made the sound decision of joining the

26. Now there was also another chemotherapeutic problem in India. Because the drug was available but scarce, there was a tendency for people not to take the full twenty-day course of injections. Feeling better, they would cease treatment and save the remaining drug for any future illness. This practice of undertreatment was not only non-curative but, importantly, caused the parasite to develop a drug resistance to antimony. These antimony-resistant strains were transmitted by the sandfly to new victims. The only other therapeutic is pentamidine, a toxic drug that costs about $100 for enough to accomplish cure. It is estimated that about 10 percent of all cases in India are now antimony-resistant. Pentamidine is the drug used to treat AIDS patients dying of pneumonia caused by the opportunistic parasite *Pneumocystis carinii*.

Southeast Asian regional office of the World Health Organization in Delhi. He was assigned as the WHO malaria adviser to the government of Burma, where he did what he could. Even the wisest of counsel couldn't make much of an impact on a malaria-endemic setting in a country whose government is beset by rebels from all quarters and effectively controls less than half of the nation. To add to his difficulties, Patnayak was unwelcome in Burma. Burmese don't like Indians, even "internationalized" Indians of high professional repute.[27] Patnayak was therefore returned to the Delhi office, where he cast about for something really useful to do. Kala azar lay waiting.

The World Health Organization, especially at the regional level, rarely has the dollars to undertake projects such as Patnayak envisaged. That kind of money usually comes from the United Nations Development Program (UNDP), which will sponsor deserving Third World projects. Patnayak got out his pocket calculator and figured how much it would cost to buy antimonial drug, wholesale, for all the estimated number of kala azar patients. That came to about $100,000. Then he figured the cost of training, meetings, and consultants—all essential budgetary items for any international-type project. That was another $100,000. Then, Patnayak thought that it would be useful to fund some applied research, such as more thorough epidemiological

27. During the British Raj, Burma was included, for administrative purposes, as part of India. The British sent numerous Indians to Burma to act as administrative officers and magistrates. As you can imagine, this caused deep, unforgiving resentment (against the Indian pawns but not against the British masters). After the British rule ended, Indian businessmen, many whose families, had been in Burma for several generations, were seen as controlling the economic life of the country. Not long after Burma gained independence, the Burmese committed the unconscionable act of expelling their citizens of Indian origin. The Burmese hold a long grudge; they still don't like Indians.

studies, serology as a means of diagnosis, and the biology of the sandfly vector—a third $100,000. The UNDP was generous: the requested $300,000 was granted, and to indicate the intent, it was designated as being under the Development Program's sector of "Prophylactic and Therapeutic Substances."

With the money in the bank, the World Health Organization convened a meeting in Delhi of representatives from India, Bangladesh, and Nepal. The WHO official in charge of the project was not Patnayak but none other than Dr. A. K. M. Kafiluddin, the former director of NIPSOM. He had left that institution a year before to become head of the Parasitic Diseases Division of the World Health Organization's regional office in Delhi.

The meeting began with Kafiluddin's review of the problem as he saw it. When he concluded, the bewildered group sat in polite silence. He had been talking about the wrong disease, another infection that he had confused with kala azar. After these niceties, the real pros, the country delegates, took over. The first order of business was the decision that it would be a winner-take-all arrangement with money going to only one institution of each country. The winners were the National Institute of Communicable Diseases of India; NIPSOM of Bangladesh; and, because Nepal didn't have any research establishment, the money for that country went to the veterinarian-epidemiologist who said he would carry out the mandated activities of the project. Then it was decided that all of those mandated activities were to be concerned with research and training. There was to be no support for existing health services to allow them to buy anti-kala azar drugs, undertake anti-vector operations, or assign more physicians to take care of kala azar patients. There had been no mention of support for

"Prophylactic and Therapeutic Substances."

During the following year there were meetings: physicians were assembled, given per diem, and instructed in the clinical ways of kala azar. Supplies were ordered and the consultants made their rounds.[28] The project period expired; the experts departed, the equipment lay fallow, and the vehicles began their deteriorative decline. All was as before, but the consumption of $300,000 for research gave the illusion that something had been done for all the Susheelas and their children dying of kala azar.

The World Health Organization / UNDP project was not, of course, the first research initiative on kala azar. The earlier work, culminating in the discoveries of the causative organism and its sandfly vector, has been described in the previous chapter. Research did not stop when the British lowered the Union Jack from the flagpole in Delhi. Studies of indigenous tropical diseases, including kala azar, continued in India, and the new government went on to support a broad-based medical research establishment. Naturally, kala azar was given low priority during those "recessionary" years when the disease had been so greatly reduced in endemicity by the vector-control operations of the National Malaria Eradication Program. However, with the resurgence of kala azar from its epicenter in Bihar, the Indian government responded in timely fashion and created, in 1975, a kala azar unit within the National Institute of Communicable Diseases. That same year the Institute established a kala azar subunit in Patna, the capital of Bihar.

28. One consultant, a member of the Indian Parliament, a politically powerful man who was a physician at one time active in conducting chemotherapeutic studies on kala azar, had the Indian equivalent of chutzpa to revise the WHO / UNDP project plan document. His revision gave him and a relative (a surgeon) a research grant of a sizable sum of money.

In keeping with the peculiar tradition of mistaken zoonotic identity, the good Dr. Chakravarty, a veterinarian, was put in charge of the Patna kala azar subunit. To assist him, a physician was also posted to the unit. This doctor, one also to be touched by tragic circumstance, was trained as an obstetrician but requested the kala azar posting because his home was in Bihar. He was a lonely man, a Christian from a minority tribe awash in a sea of socially stratified Hindus. His final burden was that not long after his arrival in Patna his wife died—or, more accurately, was allowed to die. She had fallen from the second-story balcony of their home. Unconscious, in critical condition, she was rushed to the government hospital for the necessary specialist evaluation and treatment. That day, all government physicians had gone on strike to demand higher pay. The streets leading to the hospital were chaotic, barricaded by the furious doctors. They refused to provide any service; the government hospitals were totally without medical care. The striking doctors would of course see paying patients at their private clinics, but this woman's husband, a government doctor who would not join the strike, was morally indignant to a fault. He claimed he had no resources to send her to a private clinic, refused all offers of assistance, maintaining that it was the obligation of the government doctors to take care of her. Alone, comatose in a hospital abandoned by doctors, she died.

During its first years, the Patna kala azar unit carried out some sound investigations and made a valuable contribution in characterizing the extent of the new outbreak. The staff followed several model-sentinel villages for consecutive years to determine the number of new cases each year and whether any particular age group or sex was at special risk (children of both sexes were found to have the highest infection rate). The unit was able to obtain a supply of antimonial drugs and did great service by treating people at the study village and all those who came to the Patna office.

It was also discovered that an unusual post-treatment clinical phenomenon—post-kala azar dermal leishmaniasis—was occurring in 10 to 20 percent of the cases one to two years after completion of drug therapy. Post-kala azar dermal leishmaniasis is a peculiar, little-researched condition. Its underlying mechanism is not understood, nor is it known why it develops in only a relatively small proportion of treated patients. In the normal, satisfactory course of antimony therapy, the *Leishmania donovani* parasites are completely eliminated from their normal residence, the macrophages of the deep organs—the spleen, liver, and bone marrow. These patients are cured. However, in post-kala azar dermal leishmaniasis, the parasites seem to mutate somehow to become exclusive parasites of the macrophages of the skin. The first sign is depigmented patches of skin; then the skin thickens, and finally great wartlike growths, similar to those of patients with lepromatous leprosy, develop. In fact, this resemblance to leprosy is so close that amongst those confined in India's leprosariums there may be a number of misdiagnosed cases of post-kala azar leishmaniasis. It is an extremely difficult condition to treat, not responding well to the conventional, prolonged course of antimony gluconate injections. Even more important, it is believed that these cases of post-kala azar dermal leishmaniasis act as the source of infection to the sandfly and thus are responsible for the perpetuation of the disease through the inter-epidemic years.

By the customary process of tropical laboratory entropy, the good work of the Patna unit more or less came to an end by 1980. The equipment fell into disrepair; supplies were not forthcoming from Delhi; the vehicles turned temperamental; the staff became intellectual-professional isolates. They had few journals, and few "experts" from Delhi came to work in Patna even on short assignment. The physician was despondent over the death of his wife. The Patna staff had hoped that the World Health Organization / UNDP kala azar project would reinvigorate their unit—they were, after all, *the* kala azar unit of the Institute which had captured all of the project's funds for India. This was not to be. The funds remained in Delhi because if you were working in Delhi you were *ipso facto* an important person doing important work. There was, of course, no kala azar in Delhi.

My account may appear to be a litany of complaint against research; of resources purloined in the cause of science. But I would not argue against the need for continuing research to resolve kala azar's persisting mysteries. This could best be accomplished by a Two World dualism: India has the disease and the West has the biotechnology (and the money to support that technology). The collaboration between the Indian and the Western scientist should be a natural union. The trouble is that the Indians don't like Western scientists; and they especially dislike American scientists. Animosity notwithstanding, Indian students and scientists almost to a man (there are few women in the establishment) want to study / emigrate to the United States. When you suggest that they should study at Patrice Lumumba University in Moscow because of India's special and cordial relations with Russia, they look at you as if you were a lunatic. This undercurrent of animosity goes back at least twenty years and is exemplified by an important incident that has bearing on collaborative kala azar research.

The story begins in Germany. In 1967, a medical entomologist, H. Laven, of Mainz University, published a startling report in the British journal *Nature* that when he mated male *Culex fatigans*[29] mosquitoes from Paris to female *Culex fatigans* from India, the females were rendered sterile. It was claimed further that he had completely eradicated *Culex fatigans* from a village outside Rangoon, Burma, by releasing 5000 French males (mosquitoes) each day. Laven attributed this to genetically controlled cytoplasmic incompatability. The *Culex fatigans* from all parts of the world may look physically alike, he hypothesized, but in actuality, genetically distinct strains had arisen in widely separated regions. When these genetically disparate strains were mated, the germ-line cells (egg and sperm) were so different that either they couldn't fuse-fertilize or, if fertilization did take place, the offspring—horse and donkey to mule-fashion—would be sterile.

Laven suggested that the introduction of cytoplasmic incompatibles was a potentially powerful new way of accomplishing vector population control; what could be done with *Culex fatigans* could, theoretically, be done with anopheline mosquito vectors of malaria and sandfly vectors of kala azar. This was an exciting discovery that came at just the right time. Vector control by insecticides was losing effectiveness and favor. A new, non-chemical method to attack the blood-sucking, disease-carrying insects was desperately needed. The World Health Organization, under

29. *Culex fatigans*, now more frequently referred to as *Culex quinquifasciatus*, is a ubiquitous cosmopolitan mosquito. It breeds in the most frightful muck—pit latrines, open drains, all sorts of collections of stagnant water. It is of medical importance as the intermediate host, the carrier-transmitter, of filariasis (the cause of elephantiasis) in tropical cities. The great growth of tropical cities without commensurate sanitation has led to the uncontrollable breeding of *Culex fatigans* and the increase of filarial infections.

pressure from the failing Global Eradication Program for Malaria, grasped at this new concept of vector control. WHO was also aware that a new field trial was necessary. The Indian Medical Research Council made a strong representation that the study be done in its country, and in this it had the advocacy of an Indian entomologist on the staff of the World Health Organization's Division of Vector and Biological Control. India won the trial, and by 1972 the World Health Organization had established a laboratory in Delhi staffed by an international group of entomologists—Indians, Germans, Japanese, British, Americans. Many of the American entomologists came on assignment from the U.S. Department of Agriculture since it was hoped that this type of genetic selection, introducing a lethal gene, might be applied to the many American insect agricultural pests.

The laboratory quickly got down to the work of breeding an enormous colony of Laven's kiss-of-death *Culex*. Millions of these mosquitoes would have to be released if they were to have an impact on the huge natural population. At the peak of its activity the World Health Organization mosquito ranch in Delhi was producing 5 million French *Culex* a week. Having bred the numbers, WHO selected a village several miles out of Delhi for the first field trial. What it did next was both foolish and arrogant.

Imagine a small, poor village a few miles from Delhi (there are no rich villages a few miles from Delhi or any place else in India). The houses are mud-walled and thatch-roofed. The streets are dirt and dust. There are pit latrines and a single, standing water pipe to serve the entire village. The people are mostly impoverished farmers; a few travel to Delhi, where they work as unskilled day laborers.

Most are illiterate. On a morning in 1975, a van bearing the blue-and-white logo of the World Health Organization on the door—a snake caduceus through a global map—drives into the village center. The villagers, who have a fear and loathing of snakes, regard the serpent van suspiciously. They begin to be even more suspicious when a peculiar collection of men emerges from the van—a few undoubted Indians, some strange Orientals, and some very white white men. An angry murmur of astonishment passes through the gathered group of villages when these men remove large mesh-covered cages from the vehicle, open the cages—and out flies a cloud of mosquitoes. Without a word of explanation, the snake and mosquito men then return to their vehicle and drive away. Several weeks later, the snake van appears again in the village and once more the strange foreigners release a cloud of mosquitoes from the cages. The crowd reacts—chasing the men into the van, which makes a hurried escape. A month or so later the vehicle appears again. The villagers burn it.

Clearly, the World Health Organization *Culex fatigans* team had a public relations problem. Later, they employed anthropologist apologists to explain to the villagers what was being done in an effort to gain community support, but it was too late. The village peasants may have been illiterate but they certainly weren't apolitical. During the 1970s, the Communist Party of India had considerable influence and representation at national and state levels. The villagers complained to the Communist parliamentarians, who then singled out the American scientists, and only the Americans, for vituperative condemnation. They loudly proclaimed in their Parliament and to the press that the American entomologists were really CIA agents and the mosquito project was really an experiment in biological warfare in which disease-carrying insects were being released on Indian "guinea pig" villagers.

This was the Vietnam War era; a story like this, as fantastic as it might now seem, would be widely accepted and circulated as fact. The Indian Medical Research Council which had so ardently supplicated to have the project in India maintained a pusillanimous silence. Not a word was spoken in defense. The scheme collapsed and the laboratory closed. The undercurrent of this false and vicious notion has not been dispelled by time. When the Bhopal tragedy occurred, the Hindustani *Times,* a paper regarded as the unofficial voice of government sentiment, carried an editorial saying that like the earlier mosquito-breeding scheme, this too was an example of a biological warfare experiment by the Americans against the Indians. And Laven and his sexually incompatible French mosquitoes? An American entomologist, Dr. Ralph Barr, took a closer look at Laven's data and concluded that it made no genetic sense. He then took a closer look at the mosquitoes themselves, and found that it wasn't a bad gene they harbored but that they were carrying a microbe, a rickettsia, that made the females sterile. Those French males were carrying a sexually transmitted disease to those Indian females (mosquitoes).

By 1985 it was evident that the money invested in entomological and parasitological research on kala azar had not yielded a dividend. The endemicity of the disease persisted unabated. There were no new therapies and the existing therapies were not being adequately applied. No new control projects or new research funding were proposed for the affected Asian countries. Kala azar now "stagnates," forgotten by the health ministries, who await the arrival of a new, simple, effective, and inexpensive cure.

Kala azar is not a Stars and Stripes disease like cancer, coronaries, stroke, and allergies. No American president is going to introduce a bill for Congress to fund a War on Visceral Leishmaniasis. Nevertheless, kala azar has been getting modest but steady contributions from the American taxpayer over the years. My perusal of the government records was somewhat hurried, but as far as I could discern, from 1979 to 1986 the National Institutes of Health awarded twenty-nine extramural grants (mostly to university researchers) for investigations on *Leishmania donovani*. Considering that these were peak years of the Indian epidemic (and continuing infections elsewhere in the world) and that probably one half the grants were renewals (I totaled by yearly awards) and were mostly for less than $100,000 a year (hired help is not cheap; a postdoctoral fellow will cost about $40,000 in salary and fringe benefits), not all *that* much money was being spent on *Leishmania donovani*. Moreover, hardly any of the research money went for studies on the disease in humans. There were two epidemiology studies, one in Brazil and one in war-torn Sudan, in which kala azar was one of a number of infections investigated. Nothing was funded by the National Institutes of Health for kala azar in the Indian subcontinent. Nothing was funded for study of the pathogenic mechanisms in humans and logical methods of treatment. Nothing was funded to look at the social and cultural factors that affect disease transmission. Chemotherapy got short shrift: one study on antimony treatment in mice, one study on antimony treatment in humans. Not exactly a balanced program of research on a disease of humans.

In addition to the extramurally funded research, the Laboratory of Parasitic Diseases of the National Institutes of Health also has a long history of research on the parasite.

In recent years the Parasitic Diseases Laboratory has become fashionably mainstream; its staff mainly conducts intellectualized stay-at-the-bench research. There is little of the former balance of field studies that had been a prominent part of their earlier activities. One National Institutes of Health scientist, a highly accomplished young man, did make contact with the real world by carrying out applied immunological research on kala azar in India. Within a year of beginning work, he had made a critical discovery that indicated the presence of a large, asymptomatic "carrier" population in the endemic areas. The generally accepted view had been that if you became infected, you inevitably developed the disease—kala azar. Now, immunological evidence pointed to a large group of people who became infected but had the immune wherewithal to resolve or contain the disease without ever becoming symptomatic. It was hypothesized that these "resolvers" might be the real inter-epidemic reservoirs of the parasite rather than those with post-kala azar dermal leishmaniasis. However, before our intrepid American researcher could carry out the conclusive studies needed for final proof, he made contact with another facet of Indian collaborative research reality. The host medical research institute said he wasn't giving enough credit to his Indian associates; that he wasn't leaving enough of the equipment and supplies for them to use; that he wasn't sensitive to the cultural sensibilities of the people in his study villages (the Indian "guinea pig" allusion). He was discouraged from returning to complete the studies. And so he left, pulling up the bridge of collaboration behind him.

Also in America, the Army through its Walter Reed Army Institute of Research conducts research on kala azar. I have no wish to give aid or comfort to our nation's adversaries, but if I were a spook for the other side I would keep a careful watch on what parasites attract military research interest. Realpolitik might better be reflected by the infectious disease research of the military than by the words of diplomats. Not long ago, the trypanosomes causing African sleeping sickness were singled for special scrutiny by Army researchers. Now, focus has shifted to *Leishmania donovani*—a parasite endemic in the Middle East.

So what were all these moneyed American scientists doing with *Leishmania donovani?* And if they're so smart, why is kala azar still treated by the unsatisfactory methods of the 1930s? Well, the Army researchers and their funds have been influenced by the fact of strategic life that military operations in a region of endemic visceral leishmaniasis will require new and better drugs to protect and treat the troops. In concentrating on chemotherapeutic research, the Walter Reed group has come up with some promising new therapies and novel ways to deliver the drugs to the parasite-infected macrophages. Most of these compounds are still in the preclinical testing stage in experimental animals, and they look expensive to manufacture. However, the $100 toilet seat is emblematic of military economics; cost would not be a factor in treating American soldiers. But for the Susheelas of the world, the rural and urban poor who are kala azar's real constituency, $15 for antimony gluconate is beyond their economic means.

Chapter 7

Was It Kala Azar That Killed the Dinosaur?

MOST DISEASES have their theologian-scientists; kala azar's theologians ponder the nature of the *Leishmania*'s creation and its place in its universes of sandfly and mammal. The speculation on the origins of the species makes for an intriguing story. Biological dogma maintains that parasites have evolved from free-living forms who through opportunity, mutation, and selection have come to live in or on another living organism—the "host." The host gives shelter and shops for food; indeed, for the parasite, the host *is* the supermarket. This dietary allusion may not be overly strained when we consider the Greek origin of the word *parasite*, "one who eats at the table of another."

A biologist, Lynn Margulis, has constructed the intriguing hypothesis that the *Leishmania* not only evolved to the parasitic way of life but also were *assembled* into a parasite from three different, distinct microorganisms. The Margulis theory proposes that about a billion years ago, when there were only primitive microorganisms and maybe a few sponges in the steamy soup of a young world, the first hunter-gatherer, an amoeba, ingested a yeast, or yeastlike particle of life. For countless millions of years these yeastlike microorganisms had been part of the amoeba's diet, but on that day this yeast was indigestible to this amoeba.

Life was good for the yeast inside the amoeba, and it stayed with its host and began to divide when the amoeba divided. In time, it not only became incorporated as a permanent "structure" of the amoeba but also evolved to become an essential contributor to the partnership's life process. It changed into a molecular structure (the mitochondrion), producing the enzymes responsible for the energy released by the chemical combustion of ingested food.

The final act of assembly in the Margulis theory takes place with the joining of the locomotion partner to make the one-in-three flagellated protozoan. It is proposed that simple, cork screw-like, free-living microorganisms resembling the spirochaete of syphilis were spiraling their way through the soupy sea at that time. One of these proto-spirochetes, perhaps, partially penetrated the amoeba-yeast corporation. It too was indigestible, got stuck, and became a permanent, fixed part that retained its lashing, spiraling motion. The amoeba now had a propeller; it no longer had to ooze about. It was now a flagellated organism (so the theory goes), looking not unlike the flagellated stage of the *Leishmania* (the promastigote) living within the sandfly.

It may have looked like a promastigote and swum like a promastigote but, in the leishmanial sense, it wasn't a promastigote. Evolution hadn't gotten around to that transformation because a billion years ago there were no sandflies (or any other insects). Over the slow passage of evolutionary time, life on earth progressively diversified to include forms of ever-increasing complexity. Plants became vascularized and woody. Animals acquired specialized organs and, eventually, backbones. An early life form "bit the apple"—and then there was sex. Our one-in-three promastigote-to-be spread into these new life-ecological niches, and one version came to inhabit the milky sap of certain higher plants.

These protozoa can still be found in those plants.

About 400 million years ago, primitive insects made their debut and underwent rapid (by evolutionary time standards) proliferation and diversification. Two hundred million years after the appearance of the insects, there were jumbo dragonflies with two-foot wingspreads, and the cockroach, then four inches long, settled in for its long stay. The insects radiated to every ecological nook and cranny and acquired a wide diversity of life styles. Somewhat like the finches of the Galapagos described by Darwin, it was the anatomical modification of the mouthparts that allowed the specialized life.

Sandflies belong to the Diptera—the flies. This is a large group that includes primitive free-living species, houseflies, tsetse flies, gnats, midges, and mosquitoes. Early in dipteran evolution some species developed the habit of sucking up their sustenance, and to do this their mouthparts became modified (by Darwinian selection of the best suckers) into a "syringe-needle" penetrating tube for this purpose. Some sucked up plant juices and in so doing would have ingested the promastigote-like protozoan. Our plot would have it that an adaptable mutant protozoan made a home in the fly's abdomen and became a symbiont of the Diptera. Next, the midge/sandfly ancestor would begin an occasional predatory feeding habit, first attacking other insects and sucking their vital juices, and then becoming a blood feeder with the capability of alternating to a vegetarian diet.[30] With all these new habits, the protozoan continued to dwell in the emerging species of sandflies.

30. Feeding on blood makes good nutritional sense, particularly for the female insect. It provides a diet rich in protein, the food element that will allow for the building of lots of eggs and a long life as an adult to lay them.

Now let us allow our imaginative speculation to raise the curtain on leishmaniasis. The time: 100 million years ago. Our time capsule, the fossil record, shows that there were several species of true sandflies in Europe and North America. These sandflies would have had the still non-parasitic protozoan in their digestive tract and the female fly would be taking blood meals. But what was she feeding on? Mammals, let alone humans, were a long, long time in the future. One hundred million years ago, the sandflies would be feeding on reptiles. Great big reptiles. Dinosaurs. The *dramatis personae* of our theoretical scenario are now all on stage: the dinosaur, the *Phlebotomus* sandfly feeding on the dinosaur, and the symbiotic flagellate protozoan resident in the sandfly's intestinal tract and mouthparts. When the sandfly stuck its proboscis into the dinosaur and began to feed, some of the protozoans would have been injected into the animal. The macrophages of the reptile's immune system (reptiles can make antibodies and they have macrophages) would have recognized the protozoan as something foreign, and ingested and destroyed them.

The progression of our theoretical scenario again calls upon the indigestibility principle of evolution. I propose that at some unknown point in time, probably during the Jurassic age, a strain of the flagellate protozoan had mutated in a way that made it resistant to destruction-digestion by the macrophage. Once it became impervious to the macrophage, it found the intracellular life to be nurturing—it was an advantageous character to have. And so the free-living symbiont of the sandfly became a parasite of the dinosaurs. However, its relatively large size and the extruding flagellum were not advantageous attributes for the intracellular life and, in time, selection produced a small, non-flagellate *Leishmania* form. But when the sandfly bit the dinosaur, it ingested some of the *Leishmania*-infected macrophages. The parasites were released within the insect's digestive tract, and in a manner that is still not understood, they received cues to tell their genes that they were again in their ancestral milieu and should transform to the original flagellate form. In this (theoretical) way the sandfly became the transmitting vector of the *Leishmania*.

The elegant techniques of modern genetic biotechnology support this theoretical scenario. Lizard *Leishmania* lives. Species of the protozoan continue to parasitize present-day reptiles. In attempting to define the relationship of parasites of reptiles and mammals, it is now possible to construct an evolutionary tree by a method known as DNA blot probing. It works this way: DNA is extracted from the cells—no big technical problem. The strand of DNA is then digested into fragments by a panel of enzymes, each of which will "snip" off a piece of the strand at a specific place. The bits of DNA are then separated in a gel through which an electric current is passed (electrophoresis). Next, the genes are mapped by applying a substance that the gene codes for—the probe. The probe binds to its specific locus and can be visualized. Using a number of probes and seeing if they have the same locus on the electrophoresed segments of DNA, it is possible to make a map of the genes on the DNA strand. It is also possible to do this for related species and to determine the extent of homology and difference. By various calculations which I won't burden you with, this has led to the formula that a 10 percent difference in genetic homology between related species would indicate that they had separated from a common ancestor 10 million years ago.

In 1987, S. M. Beverly and his colleagues applied this DNA hybridization method to construct an evolutionary tree for the *Leishmania*. Their study of the species and subspecies of *Leishmania* in humans and the species in reptiles led them to conclude that all *Leishmania* arose from a common ancestor 80 million years ago, a time when the dinosaurs walked the earth. Which brings me to my own Dinosaur Apocalypse theory. Each science to its own apocalypse. The paleontologist would have you believe it was the Big Bang from a meteorite smacking into our planet, an event that caused a worldwide dust cloud to occlude the sunlight. The suddenly cool world could not support the life needs of the dinosaurs and they all died. Maybe. Consider an alternate explanation by this parasitologist.

We know from the fossil records and interpretation of DNA homology analysis (a kind of fossil record from the living) that (a) there were *Phlebotomus* sandflies during the age of dinosaurs and (b) the *Leishmania* had their origin during that remote time and became parasites of those reptiles. Could it be that the dinosaurs were immunologically inexperienced to this new parasite and that the *Leishmania* were highly virulent in those animals? Did leishmaniasis, a new disease, sweep through the dinosaur population as AIDS is now doing to the vulnerable human population groups? Was it an epidemic of reptilian kala azar that caused the extinction of the dinosaur?

Apart from these amusing exercises in evolutionary speculation, the problem that has most piqued the curiosity of research workers, and those who fund them, is the problem of how the *Leishmania* manage to live within the macrophage without being destroyed. Normally, macrophages do their killing by several methods; they manufacture "globules" of lytic enzymes (lysosomal enzymes), which coalesce into a vacuole surrounding the ingested (phagocytized) microorganism. In addition to the digestive lysozyme, the macrophage kills by making superoxides, such as hydrogen peroxide, as a product of a respiratory burst of glucose metabolism. What makes this all even more complicated is that there are macrophages and there are *angry* macrophages. The run-of-the-mill macrophage is non-specifically omnivorous; it will recognize anything that is foreign, such as microorganisms, and phagocytize it. It does this in a relatively sluggish and inefficient manner. However, after an infection with a microbial pathogen or after an effective immunization, the macrophage will have received powerful chemical signals (lymphokines) from a turned-on T lymphocyte to make it a highly specific and highly effective "contract" killer. Now, when it encounters that microbe, and only that microbe, it will rapidly engulf it and produce the enzymes and superoxides for a quick kill. But the macrophages of people with kala azar do not become angry; the *Leishmania donovani* continue to live and multiply within them.

Recent research on humans with kala azar and experimental animals infected with *Leishmania donovani* has revealed the versatility of the parasite in confounding the macrophage. Three evading strategies have been described. First, the *Leishmania* has acquired the ability to bind on to itself a naturally occurring protein in serum known as complement (C3). For some as yet unknown reason, the C3-coated *Leishmania* inhibits the respiratory burst, which results in the production of the killing superoxides. In general, the oxidative metabolism of the infected phagocyte is impaired.

The second mechanism to account for the *Leishmania*-infected macrophage's failure to become angry (along with a general immunodepression of the cell-mediated immune system—as in AIDS) is the parasite's ability to shut down the required communication between the specialized cells of the immune system. The membrane around each cell of the body bears a molecular code that is personal (except in the case of identical twins). These are the histocompatibility antigens (in immunological shorthand, the MHC and HLA antigens), which are best known as being responsible for rejection of transplanted organs. If the macrophage is to be activated, to become "angry," it must receive the lymphokine signal from the T cell; and to make that lymphokine, the T cell must first contact the macrophage to ascertain that it shares the same histocompatibility type. It's not going to turn on for some outsider—this is a family affair. One step further: the histocompatibility antigens are proteins, and like all proteins their assembly is directed from the coded instruction of a gene—the MHC genes in this case. *Leishmania donovani* has now been shown to inhibit the expression of the MHC genes (for you science buffs, by increasing intracellular prostaglandin E_2), and so the infected macrophages bear few identifying histocompatibility antigens on their surface for the T cells to recognize. Very complex. Very cunning.

The third survival mechanism is also mighty shrewd. By a means yet to be elucidated, the parasite stimulates the bone marrow to produce more and more macrophages. The assembly line is so active that it doesn't make a finished product—immature macrophages are elaborated into the immune system. These immature macrophages can capture and engulf the *Leishmania* but are physiologically incapable of killing them, even in the presence of stimulating lymphokines.

In the National Institutes of Health grant application form, there is a section in which the applicant must state how the proposed research will be of actual benefit to human health. No matter how remote the study is from the human condition or how abstruse it is in objective terms, there are always "connections" that can be made to satisfy the reviewing peers, who are mostly benchpersons themselves. There are good, unimpeachable words that can be used, and two of the best are "vaccine" and "immunization." There has never been a vaccine to protect or cure *any* parasitic disease of humans, but the search goes on. For the various forms of leishmaniasis, including kala azar, the focus and funding has been on the *lack* of functional immunity in the infected host and how to stimulate the immune system to better effect.

Kala azar patients produce abundant antibody against the parasite—diagnosis can even be made by detecting the antibody in a patient's serum. The antibody seems to have little or no effect on the parasite, probably because it can't get inside the macrophage where the *Leishmania* enjoys the privilege of its sheltered position. To kill the *Leishmania*, the other immune arm, cell-mediated immunity, has to be activated. This arm of the immune system, presided over by the specialized lymphocytes—the various types of T cells—"enrages" the macrophages in the way I have just described and has a T-cell subset, the killer cells that directly give the kiss of death to the unwanted immigrant by means that do not involve antibody. The macrophages of kala azar patients, as already noted, are not much better than a "dead leech"—as my grandmother used to generically characterize all the animate and inanimate dysfunctional. It's not only the macrophage; studies in mice and men infected with *Leishmania donovani* have revealed that the lymphocytes are also not much better than my grandmother's dead leech. They not only aren't "turned on" but can't be activated by the test tube techniques that normally do so. The reason for this cell-mediated immune indolence, this anergy, is not known, although Dr. David Wyler of Tufts University has found a factor in the serum of kala azar patients that suppresses T-cell activation.

Whatever the underlying cause or causes of this anergy may be, it is obvious that immunological cure cannot proceed unless the cell-mediated arm is, somehow, made to switch on. Immunotherapy has been one experimental approach to do this. Lymphokines—which biotechnological recombinant gene methods are providing in ever-increasing numbers in pure form—have been injected into experimental animals with encouraging, but not spectacular, effect. Brazilian workers have carried out the only trial of immunotherapy in humans with leishmaniasis. This was not on kala azar but in patients with the muco-cutaneous form caused by *Leishmania braziliensis*, a very nasty disease present in tropical South America. They inoculated an extract from lymphocytes made from people who had been drug-cured of the infection. It was reported to give better-than-drug effect in people who had recently acquired the infection, but had no effect on patients with advanced disease. Nor have conventional vaccines of dead or inactivated *Leishmania donovani* ever been successful in immunizing against kala azar. Recent research has exploited biotechnological techniques in isolating the many antigens that comprise the parasite. The aim is to find an antigen which will specifically activate the cell-mediated immune arm and be protective as a vaccine. Not much luck there, either.

One cannot fail to be impressed and excited by the vitality and innovativeness of this large and diverse body of recent research on the leishmaniases. Without doubt the new knowledge emanating from the application of elegant biotechnological methodologies delights the intellect of all students of host-parasite relationships. Yet my thoughts keep returning to the primary health centers where Susheela failed to obtain drug for her sick child and where Dr. Nurul Islam Khan was examining bone-marrow samples through a clouded microscope in a room lit only by sunlight filtering through an open window. There is too great a gap between research and reality; between the high intellectual world of biomedical research and the needs of the sick and those who attend them.

II

Malaria: From Quinine to the Vaccine

Chapter 8

In Another Village a Mother Dies

ON THE BUFFALO RIVER—the River Kwai—at the town of Kanchanaburi, the big familiar relic, the bridge, still remains; it is an unremarkable structure spanning the river to nowhere (shortly after the war the Thais tore up the tracks to the Burmese border). The festive, touristy scene around the bridge—souvenir stands, food stalls, touts for boats with bongo drums—is an unseemly commercial celebration of the unspeakable atrocities committed here. In what year will T-shirts be sold in tourist shops of Auschwitz? The ultimate relics, the bones of the fallen and felled, lie in the immaculate graveyard of the British and in the quiet, more jungly cemetery of the Dutch and Australians.

One hundred miles to the west, the Kwai begins its flow from the jungle-covered, not-quite-mountain hills at the Burmese border. A great dam has been built across the upper Kwai and a valley flooded into a lake. At the head of the lake is the new, raw town of Sangklaburi. Three Pagoda Pass is ten miles or so from Sangklaburi. It was through this pass that the Burmese armies with their "tanks"—the war elephants—spumed forth to ravish and pillage the early kingdoms of the Thais, first at Sukhothai and then at Ayuthia. Now, when you walk to the Burmese side of Three

Pagoda Pass (there are no Customs or immigration barriers at this remote border), signs and banners proclaim this to be the headquarters of the Karen and Mon Liberation Armies. The streets are patrolled by menacing child-soldiers carrying AK-47s. It is wild country, disturbed by civil strife and ecological change.

The last relics of the Kwai's Japanese rail line to oblivion lie in the schoolyard of a Thai village several roadless miles north of Three Pagoda Pass. The schoolmaster shows the visitor the remnants of the depot and the termite-ridden remains of the ties. The school assembly bell, hanging by a rope from a limb of a mango tree, is the nose cone from an unexploded bomb. Next to the school, raised on pillars, is the wooden residence of a group of monks. On this late morning in June their prayers have ended; only the unceasing anguished cries of a monk dying from throat cancer break the subdued quiet of the village. In a one-room, wood-framed, tin pan-roofed house at the village edge, Amporn Punyagaputa, twenty-three years old and big with child, sits alone, feverish and confused by the searing pain in her head.

Amporn is not an ethnic Thai. She is a woman of the hill tribes, a Karen, born and raised in a mountain village near that dodgy border between Thailand and Burma. As a child she had a few years of education in a Thai primary school located in a nearby village, but she was forced to leave when only marginally literate. The Thai officials decided that education was a privilege reserved for Thai citizens, not for the tribal peoples who had no real national affiliation. Also, when Amporn was eleven years old the family had to migrate to another part of the forest. The ten-year cycle of land use was at an end. Now it was a time to allow the exhausted plot to regenerate to nurturing forest.

It was time to move, to slash and burn a new clearing for their modest crops of maize, beans, mountain rice and—maybe—a little opium poppy.

That was when she was a child and the traditional life style of the hill tribe swidden farmer remained unchanged. A relatively stable tribal population size permitted this. The cost of that stability was a fearful childhood mortality that exceeded 40 percent. The desire of fathers to have large families balanced this wastage and ensured tribal survival. The fathers wanted large families to expand and work the plots so that more cash crops could be raised. They wanted the money, $800 to $1,000, to buy the ultimate status symbol of a Karen male—an elephant. They craved ownership of an elephant as other men yearn for fast cars, large homes, and many young wives. The ratiocination being that the elephant could be let out to hire by the timber concessionaires; but really this didn't make economic sense considering the cost of the animal's care and feeding. In truth, Karen men simply like elephants.

Increased access to, and usage of, primary health facilities gradually upset the population size so delicately balanced between a father's demands for a large family and the death of many small children. As always, the children frequently became sick, but now fewer died. Now, they could be taken to health stations staffed by nurses and medical assistants dispensing life-saving medicines to those acutely ill with diarrheas and respiratory infections. Family size and tribal populations rapidly increased. Fathers thought that the elephant was finally within their grasp. They failed to reckon on the realities of demography and the capacities of swidden agriculture. The swidden plots and the farming methods that had served so well for countless generations failed to provide the food for a population that had nearly doubled within a generation. The farmers tried to shorten the fallow time, but this only quickened the exhaustion of the land. There was less and less forest available that could be cleared for farming. The forest was being drastically reduced by timber concessionaires on behalf, mostly, of the Japanese, who wanted the Thai and Burmese forests in order to maintain the "wooden" aesthetic of their culture. Some of Amporn's family, like so many other of the close-knit Karen families, were forced to the heartbreaking necessity of leaving the tribal group to seek life and employment elsewhere. Amporn and her husband—she had been married the year before—were amongst those who made the exodus from the high hills of their birth.

Not long after their marriage they moved to this village on the Thai side of the border near Sangklaburi. Most of the villagers were also displaced Karens trying to eke a livelihood from one or two acres of land leased from Thai landlords. Amporn and her husband were wise and careful farmers, and they leased a two-acre tract in the expectation that it would provide sufficient food and cash crops to meet their modest needs and, eventually, give them enough savings to buy their "elephant"—a television set. But the Thai owner had extracted so heavy a lease rent that Amporn's husband had to supplement their income by working for the sugar estates bordering the lower Kwai seventy-five miles away. Amporn was now alone most of the time. She bore the full responsibility of the farm. And she had conceived four months after her marriage.

The joy and stoicism that Karen women have in bearing their first child was diminished for Amporn. Here in this valley village she felt so alone and so unrelievedly tired. So far, she had been fortunate not to sicken with malaria as did so many of her neighbors. In the highlands of her home malaria was virtually unknown, but here it seemed to attack everyone. Many would have died if it were not for the medicine dispensed by the health center in the neighboring village. It must be the bad waters of this place. In school they tried to tell her some nonsense about mosquito bites and malaria. Every Karen knew that drinking the corrupted waters of the lowlands was the cause of malaria. When she first came to the village, a year ago, there was so much malaria and the village headman was so angry with the state of affairs that he sent water samples to the health officials in Sangklaburi, demanding that the water be analyzed for whatever it was that caused the sickness. To its credit, the government responded promptly and sent a malaria office team to the village; but these men, instead of treating the water as they should have done, sprayed all the houses in the village with an insecticide. A strange thing happened after that: most of the village cats died.[31] After the cats died, there were many more rats and people began losing a good part of their food crops to the rodents.

Two weeks ago, when the maize began to ripen, Amporn started sleeping in the shelter she and her husband had built at the edge of their fields. There were so many rats that any additional loss from marauding wild pigs and the occasional bear would bankrupt her fragile economy. During those watchful nights on guard she would clang the gong to drive the wild animals away. She slept fitfully in the shelter; the attack of mosquitoes was unrelenting. Her neighbor, an elderly lady of long residence in the village, had told her that before the loggers came to cut the forest

31. The insecticide was DDT, which has no significant toxicity for humans. Cats are more susceptible and there are numerous reports of village cats dying within one week after malaria-control teams have sprayed DDT onto household walls.

and before the dam was built and the impounded lake formed, there were not so many mosquitoes. In that former time, sleeping out in the shelters of their fields, as is customary for farmers in Thailand, was not a problem. The old lady now slept under a mosquito net and advised Amporn to do the same. A net was expensive, but Amporn thought they could afford one after the baby was born.

Amporn had been reasonably healthy during her pregnancy. She was a modern, informed woman, and last month had gone to the "mothers-to-be" clinic at the nearby health center. The nurse had given her some vitamin pills, told her the baby within her was doing fine, although Amporn, like most pregnant women in the district, was anemic and the nurse gave her more pills for that. Amporn wasn't sure what "anemic" was, but she knew it was something to do with her not having enough blood. That was probably why she felt so tired and had to drag herself through the day's work. The nurse at the clinic also told her that she should go to the hospital at Sangklaburi to have the baby. There she would have expert care and it would only cost 100 baht ($4).

Amporn may have thought of herself as a modern woman, but she was also a Karen and she had consulted the friendly old village midwife. The midwife too, after prodding Amporn's abdomen, predicted a normal delivery and advised that she be called when labor began. Even at this late stage of pregnancy Amporn was still undecided where she would give birth. However, since her husband would likely not be there to arrange transport to the hospital in Sangklaburi, she probably would have the baby at home, attended by the traditional midwife. She would also then be in familiar surroundings, amongst the people she knew, and the proper rituals would be performed to protect the newborn's soul from harm.

Amporn's world—her personal concerns and joys, the comfort of the day's domestic routine—had begun to vanish yesterday morning, submerged under a sudden wave of sickness. The child within her became an insupportable burden, her back ached, there was a nausea so intense that it made her choke breathlessly. The attack came with surprising ferocity. In a moment the nausea yielded to a chill that made Amporn feel her body was encased in a shroud of ice. Under the blazing tropical sun she shook uncontrollably. During this "freezing" rigor, Amporn's temperature had risen to 104°F. After an hour of tooth-chattering shakes the rigor abated, and for a few moments in the eye of this parasitic storm Amporn thought she might yet live. The brief respite was followed by a feverishness that was as intense as the sensation of cold she had experienced during the rigor. Amporn's temperature was now 106°F. Her senses reeled; consciousness blurred. She crawled into her house and collapsed upon the cool dirt floor, her sarong sodden with the sweat pouring from her burning body.

In the early evening, the fever broke and Amporn, exhausted, drained of all strength, fell into a fitful sleep. The cries of the dying monk woke her in the first light of the new day. Still lying on the floor, she sensed a foreboding communion of death with the tortured holy man. She was too dispirited to eat but had a great thirst. As she staggered to the large water-storage klong jar, the frightening premonitory signs of another rigor began. Amporn realized that it was malaria. She knew she needed medical help, but her weakness and confusion prevented her from even calling to her neighbor. The malaria again seized her. It was at this point we began our story: Amporn Punyagaputa, twenty-three years old and big with child, sits alone, feverish and confused by the searing pain in her head.

Now there was another terrifying aspect of the malaria attack: she felt her womb contracting. The pains of child-birth were beginning. But the baby was not due for another two months yet. What was happening? The headache became unbearable; she lost sense of time, of place, of her person; her vision blurred; she felt herself being swept into a dark vortex. Then there was nothing. Amporn lay unconscious on the hard dirt floor, her womb, propelled into premature labor by the malarial fever, starting its rhythm of birth.

Amporn's good neighbor, Lek, was also a Karen woman displaced from her hill tribe village. Most mornings Amporn and Lek would grind their grain in the stone mortar mills in front of their houses. They would exchange gossip and commiserate on their lonely lives. Lek's husband too was away a good deal of the time, working as a logger in the deep forest. Amporn's failure to appear worried Lek, and at mid-morning she decided to see if all was well at the house next door. All was not well: Lek found her friend collapsed on the floor unconscious, her clammy skin an ominous gray pallor. Lek quickly realized that Amporn had to be brought to the hospital at Sangklaburi . . . she was beyond the help of traditional medicine.

At that time of day the men were away working in the fields. Only the women were in the village to give assistance. The monks could offer their advice and prayers, but they were forbidden to touch a woman, even a dying woman. There was no formal village civil defense unit or Red Cross chapter, but over the years of too frequent emergencies—the wounds of the agriculturist, the sudden life-threatening fevers, the births gone bad—the women had developed their own efficient disaster drill. Responding to Lek's summons, the women gently placed Amporn on the canvas stretcher that had been stored in the Headman's house. Then they set off on a walk-trot, spelling each other as bearers, over the two miles of path to the rutted dirt track that served as a road to the surrounding villages.

Once at the road, the women had to wait almost a half hour before a samlor taxi putt-putted by. The samlor is a motorcycle converted to a tricycle with a bench seat for the passengers. It is an uncomprising, hard ride, but reliable and economical—the proletarian transport of Thailand. Packing a comatose, seven-month-pregnant lady and her traveling companion into a samlor with its four-foot bench required some shoving and folding. Amporn's head rested on Lek's lap; only her occasional groans gave sign that she still lived. In this manner they finally came to a jolting stop under the portico of the Sangklaburi hospital.

The portico of the country hospital is an intersection of old and new Thailand. The Sangklaburi hospital was but a few years old; it was the standard twenty-bed rural affair, of standard government-dictated architecture, plain and functional except for the portico which graced it with a pleasant elegance. A large gilt Buddha sat smiling, in bronze solidity, at one side of the portico facing the garden. Traditional and reassuring. The hospital "kitchen" was also housed under the portico—a "noodle lady" and her steaming cart. The noodle lady's pet gibbon, a furry blond baby, sat pensively in the Buddha's lap.

The nurses came running to their patient and carried Amporn across the portico into the examination room. Amporn was now in the care of the new Thailand. This small, immaculately maintained hospital in a far corner of the kingdom was remarkably well suited to serve the health needs of the local population. There was a pharmacy stocked with a broad and adequate supply of drugs and biologicals. There was an X-ray machine, an efficient laboratory, and an ambulance to take patients to the Provincial General Hospital if their problems were beyond the capabilities of the young staff and their resources.

The people of Sangklaburi, like the rural populations everywhere in the Third World, were served by the least experienced and youngest physicians. The government sent the new medical graduates to staff the rural hospitals and health centers. After a few years in the country they were brought back to the cities for specialist training, to enter private practice or to continue government service at the general hospitals. In the balance, in Thailand at least, service by the youthful was not necessarily a disservice to the sick. It requires the energy and enthusiasm of youth to be a country doctor in the Third World. And by no means are these young doctors incompetent. They receive intensive training in basic clinical medicine—family care—as part of their medical curriculum. Once posted to the rural health facility, the government does not abandon them. They are brought to the Provincial General Hospital for frequent short courses and meetings with senior specialists, who can also be relied upon for assistance with the difficult cases. The young Thai doctors are given decent housing, decent pay. Equally important, the nurses and technicians working with them are also given decent housing and decent pay. This makes for an egalitarian camaraderie, a harmony and vitality in working at full-time patient care. Nor do these kids remain inexperienced for long. They are bright and learn fast from the exigencies of daily practice. And, for what they haven't yet learned, there are the older nurses to keep the young doctors from doing too much harm.

In the examination room, the team of doctor and midwife nurses quickly came to the provisional diagnosis of premature labor brought on by acute malaria. Amporn was not an unusual case. Each week during the early rains (the height of the malaria transmission season) one or more young women, pregnant and gravely ill with malaria, were brought to the hospital. Malaria of the malignant tertian type caused by the parasite *Plasmodium falciparum* was always of concern because of the rapidity with which it could develop into a life-threatening infection; but pregnancy imposed a degree of vulnerability even beyond the fearful "natural" risk.

To be pregnant is to have within your person a growing foreign body that meets many of the biological definitions of a parasite. The mother makes all sorts of immunologically dampening adjustments so that this particular foreign body, the fetus, will not be rejected. While it is not an AIDS-like depression, the immune compromise is such as to make a pregnant woman of the tropics more likely to contract malaria and to die of the infection.[32] A woman of the Third World may face a four hundred times greater risk of dying each time she becomes pregnant than a pregnant woman of an industrialized country. In regions of malarial hyperendemicity, a high proportion of this mortality can be attributed, directly or indirectly, to falciparum malaria. Survivors tend to have a severe anemia (malaria, hook-

32. The mechanisms responsible for pregnancy-related enhanced malaria are still not fully understood. One important cause of this pathogenicity is brought about by a characteristic of the malignant tertian malarial parasite, *Plasmodium falciparum*. Red blood cells infected with this parasite congregate in the placental blood vessels—sometimes one can hardly find the parasitized erythrocytes in the "finger-stick" peripheral blood while at the same time the placental blood is, literally, thick with infected erythrocytes.

worm, and nutritional factors, all present in the pregnant peasant, can turn blood into "water") and the spontaneous abortion rate is as high as 50 percent. The babies that *are* born to the malarious usually have a birth weight 200 grams or more below normal. A low-birth-weight baby who doesn't "catch up" has a puny immune system, and these children frequently die during childhood of all sorts of infections. The death certificate (when there is one) may read diarrhea or pneumonia, but the real cause of death was malaria of the mother.

There were other fevers, other causes of coma—the virus of Japanese encephalitis and a group of bacterial pathogens capable of invading the central nervous system—that were old familiars in western Thailand. Thus, before treatment could be given, the presumptive diagnosis that Amporn's fever and coma was caused by malaria had to be confirmed. The laboratory technician pricked Amporn's unfeeling finger and from the droplet of blood she made a film on a glass microscope slide and filled a glass capillary tube. Thirty minutes later the technician returned to the group gathered about their dying patient to give them the report that more than 15 percent of Amporn's red blood cells held the tiny signet ring-like forms characteristic of *Plasmodium falciparum*. This was an intensity of infection considered to be potentially lethal and too often beyond the curative power of antimalarial therapy. The technician also reported that the reading of the centrifuge tube gave a hematocrit of 17 percent—Amporn was so anemic that she hardly had enough red blood cells left to oxygenate a mouse.

So the young doctor was confronted with a conglomerate of three emergencies requiring immediate attention. The overwhelming malaria infection had to be brought rapidly under control by a chemotherapeutic agent. The anemia, so profound as to be nearly insupportable of life, had to be reversed. And the premature labor had to be halted if the unborn child, if not already dead in the womb, was to be saved. Nor were there any simple textbook solutions to this triple threat. Twenty years ago Amporn could have been given an intravenous injection of chloroquine and within hours this remarkably efficient antimalarial drug would have begun killing the parasites. However, over the years the widespread use of chloroquine had led to the selection of resistant strains of *Plasmodium falciparum* that no longer responded to the drug.

The doctor didn't even consider using chloroquine. Instead, to treat Amporn, he opened a vial containing a solution of a plant alkaloid that had first been used to treat malaria more than 350 years ago—quinine. Quinine was all he had in his therapeutic armamentarium. It was all there was to treat severe falciparum malaria. But quinine was toxic, relatively slow-acting and, importantly in this case, tended to induce abortion. Quinine also had a side effect of stimulating the pancreas to produce more insulin. The insulin could "burn off" what sugar Amporn had left in her blood and send her into a state of shock and irreversible coma. Still, all chemotherapy, be it taking an aspirin or taking quinine, is a chancy clinical compromise. Necessity dictated that a nurse prepare the syringe to infuse the quinine slowly into Amporn's nearly collapsed vein.

Amporn urgently needed new blood. In Sangklaburi hospital there is no refrigerated blood bank; the blood for medical emergencies is stored in the original containers—on the hoof, so to speak. The laboratory identifies the patient's blood type and matches it with that of a relative from whom blood is drawn. When there are no relatives, as in Amporn's case, the hospital staff selflessly act as donors.

Amporn was found to have a common blood type shared by several of the nurses. One of those nurses, Chalong, immediately volunteered to give his blood. Chalong is a young man, the same age as Amporn, who had spent two years in a nurses' training program at a government general hospital. He loved nursing and was the first to volunteer for any job. He was the hospital's dynamo, completely and single-mindedly devoted to taking care of the ill. His smiling, calm, youthful presence seemed always there, at any time of the night or day, when needed.

A unit of blood was withdrawn from Chalong and transferred into Amporn's vein. Within minutes her color improved, her pulse was less thready, her blood pressure rose—there was yet hope that she could be saved. Later that day another unit of blood was given to Amporn, this one collected from a vivacious, slender nurse known affectionately as Gop, Thai for "the fox."

The third problem, the premature labor, posed a special dilemma. Drugs could be given in an attempt to stop the still early birth process. However, the doctor could not detect a fetal heart beat—the unborn baby was probably already dead.[33] It might even be better to encourage the abortion, although this was an ethical and legal decision the young doctor had to make each time he treated a pregnant woman suffering from severe malaria. His, and others', observation had been that if they induced abortion to remove

33. What kills the fetus when the mother has severe malaria is not known with any certitude. The malaria parasites rarely cross from the placental into the fetal circulation, so it is not a malarial infection in the unborn baby that is the cause. Probably, it is a combination of the high fever of the mother, which makes the womb a cauldron, and the deficiency in transferring oxygen from the placenta, where blood circulation is impaired by the parasite and inflammation clogs blood vessels, that leads to the death of the fetus.

the heavily parasitized, "toxic" placenta, the mother would have a much better chance of surviving, of responding to the antimalarial chemotherapy. But this was only an observation, it was not prescribed medical practice. Abortion, although frequently performed by private practitioners, is in fact illegal in Thailand. However, Amporn was already in labor, the baby was probably already dead, and it was decided to allow nature to take its quinine-accelerated course.

Amporn was taken to the women's ward, screens placed around her bed. The ward nurse was almost constantly with her. Near midnight, Amporn, still comatose, showed signs that she was about to give birth. She was quickly moved to the delivery room where, with the midwife-nurse in attendance, she gave birth to a dead premature infant. It was a boy.

In the morning, Amporn opened her eyes. The coma receded; the fever abated. The blood film now showed fewer parasites—they were still there but the quinine was working. The hematocrit had risen to 25 percent—still very anemic, but improving after the blood transfusions and the continued intravenous infusion of a glucose-saline solution. The doctor came to her bedside and tried to explain what had happened. Amporn's mind was still blurred; it was difficult for her to form thoughts, but somehow she understood that she was in a hospital and was being cared for.

Throughout the day Amporn's condition remained stable; the parasites in her blood continued to diminish. The staff's mood brightened from the somber tension that pervaded whenever they were dealing with a critically ill patient whose life was in the balance . . . Amporn might yet make it, they now thought. The Gop talked about a party she was organizing—a boat would take them on a cruise around the lake.

There would be a mountain of food, singing, swimming. Hospital activities returned to a less stressful normal; outpatients were seen by the two doctors, prescriptions filled, a fractured arm set. In the late afternoon, their busy day finished, the staff gathered in one nurse's house for a heated card game. Another nurse cooked a big pot of green curry chicken for a communal dinner.

Sometime in the very early morning hours Amporn died. Her tired heart just stopped beating. The night-duty nurse called the doctor who, still in his sleeping "sarong," came running from his nearby house to the hospital. Everything possible was done to revive Amporn. The staff worked quietly and efficiently for over a half hour to no avail. At four in the morning, feeling a great, futile depression, they left, and Amporn was taken to the morgue to rest beside her dead son.

During that night, a child was born to another Karen woman. What the personal tragedy or circumstances of this woman were will never be known, but at first light she painfully rose from her bed and fled the hospital, abandoning her newborn infant. In the morning sunlight, Chalong sat on the bench on the hospital veranda tenderly holding the baby and sobbing, the tears streaming down his young face.

Chapter 9

The M&M's: Monkey, Man, and Malaria

FOR CENTURIES distancing beyond recorded history, malaria has been a pregnant ladykiller. Malaria also kills the children; in regions of intense transmission, 40 percent of the toddlers may die of acute malaria. Malaria also kills the immunologically "unsalted" adults—migrants from teeming Third World cities who pioneer new agricultural lands, soldiers of the Western world battling to save democracy in tropical nations, tourists, businessmen. In 1990—in this age of rocket ships and genetic engineering—250 million people will get malaria and at least 2.5 million will die of the infection—needless deaths. Malaria is not an AIDS; the curative antimalarial drugs are available. Malaria is not like cancer; the most intimate details of malaria's causation are known. Malaria is not like the epidemic of drug addiction; given the resources, successful antimalarial campaigns can be implemented.

This great death is caused by so small a parasite, a protozoan of the genus *Plasmodium* acting in concert with its agent of transmission, the mosquito. Humans are not alone with their malarial parasites; almost all vertebrates from snakes on upward are parasitized with *their* species of *Plasmodium.* Between the world wars, bird malarias were extensively used to screen drugs for antimalarial activity.

The most promising and non-toxic compounds were later tested in humans. The Germans favored experimental infections in canaries, while the British preferred a malaria of chickens, but this difference probably has little sociopolitical significance. From time to time flocks of turkeys will be wiped out by *their* malaria. It is hypothesized that the immunologically naive native Hawaiian birds were killed off in great number, some species annihilated, by the avian malarias carried by the Asian birds that flocked, unregulated, into the islands during the late nineteenth and early twentieth centuries. After all, there wasn't even a *mosquito* in those once pristine isles before the *haoles* (Europeans) brought them as stowaways in the drinking-water barrels of their sailing ships.

But those malarias are for the birds, and what really is of more interest as the origin of the species, both personal and parasitic, are the malarias of our cousins the apes and monkeys. The table talk of biologists, particularly, older, semimolecular biologists, frequently turns to speculations of human origins and those of their lifelong companions, the animals and plants they study. Medical parasitologists have a package deal for such speculative exercises since people and parasites have, for the most part, evolved together. With a good vintage and good company, the early evolutionary conundrum may be argued with some passion—did the primitive malaria begin as a parasite of some prehistoric reptile[34] that later was picked up by a mosquito,

34. The hypothesis for this being that the parasite began as an invader of intestinal cells and later started to parasitize white blood cells and red blood cells. Protozoan parasites (the coccidia), related to the malaria parasites, which are confined to the intestinal epithelium, are still found in a great variety of vertebrates—including humans. One of them, *Cryptosporidium,* can cause a killing disease in AIDS patients.

or was it first a parasite of the mosquito that later became established in the reptile? However, since there are no fossil records of tissue parasites, it is a debate that will continue unresolved for the years to come.

Surely an evolutionary scenario for monkeys, man, and malaria could be more easily and logically constructed. First we could list the species of malaria parasites naturally infecting humans. Then we could look for morphologically similar analogues of those species in higher apes and successively down the primate line to the lemurs and tarsiers. To confirm the evolutionary ascendancy tree we have constructed from physical likenesses—similarity in the form and shape between species of malaria *(Plasmodium)* parasites—we could carry out cross-infection experiments. By that experimental scenario, the malaria parasites of humans should be infective to higher apes—chimpanzees, gorillas, and orangutans—but maybe not to the lower apes, the gibbons. And vice versa, the malarias of those big apes should be able to infect humans. The malaria of monkeys should, maybe, be able to infect their anthropoid ape relatives but not humans. All very tidy in theory but, unfortunately, reality confounds logic; there is a massive mishmash of confusing observations and seemingly illogical host-parasite relationships.

Some, but not all, species of malaria parasites of humans can infect the high and low apes. Some malaria parasites of apes that resemble species in humans can infect humans, while other lookalike species cannot. The "second cousins" of humans, the monkeys of Asia and Africa, are totally refractory to infection by the human malaria parasites, but a "third cousin," a monkey of South America, is susceptible. There are malaria parasites of Asian monkeys, morphologically analogous to species in humans, that will infect humans; other morphologically analogous species cannot infect humans. There are malaria parasites of monkeys with no analogue in humans that will, nevertheless, infect humans. To moil the matter even more, persuasive evidence exists of instances in which, malarially, the human has been more of a menace to the primates than the primates to the humans. There are species of *Plasmodium* in some higher apes and South American monkeys that not only *look* like malaria parasites of humans but *are* parasites of humans. These infections, established in the primate host, are thought to have been first acquired from mosquitoes carrying the *Plasmodium* of humans.

Research on primate malaria parasites has not been a pursuit merely to satisfy scientific curiosity—a parasitologist's toy. Monkey malarias have been both of great service and a potential threat to humans. The connection between monkey malaria and benefit to humans is syphilis.

The pre-World War I period, as now, was one of intellectual, artistic, and sexual adventure. It was also a period, as now, when syphilis was on the increase. In 1905, Fritz Schaudinn in Germany discovered the threadlike, corkscrew microorganism in the blood, brain, and tissues of syphilitics that he named *Treponema pallidum.*[35] As is the case for many of our medical problems, knowledge of cause does not automatically bring immediate knowledge of cure (prevention, maybe, but not cure). It was not until 1912 that Paul Ehrlich ushered in the science of chemotherapy with his synthesis of "606," salvarsan, the organic arsenical

35. Microbiologists in those days were cross-workers, and pathogens of all kinds came under their purview. Later, Schaudinn was to work on the life cycle of malaria and he really screwed things up.

which was the first effective therapy against syphilis.[36] Salvarsan now has, of course, been replaced by penicillin as the treatment of choice. However, neither salvarsan nor penicillin gives satisfactory results for the treatment of late-stage syphilis, when the pathogens have made their way to the brain and other parts of the central nervous system.[37] The nature of the blood supply to the central nervous system is such that drugs do not readily cross from the blood, through the vessels, into the nervous tissue—the blood-brain barrier.

One may think of pre-World War I Vienna as the city of Strauss and strudel, but on the darker side its asylums and neurological clinics were filled with syphilitic paretics. These were men and women made crazy by the corkscrew microbes in their brains. Some were blinded, some paralyzed, many throwing their legs out in the ataxic, tabetic gait of the late-stage syphilitic. Ibsen's play *Ghosts* is as much a story of the pathetic plight of the paretic as it is a feminist argument. The play ends with young Oswald Alv-

36. The organic dye industry which was beginning to flourish in late nineteenth-century Germany was the springboard to chemotherapy. Ehrlich noted that certain dyes would stain—combine with—only certain microbes. This led Ehrlich to the idea that these dyes would act as "magic bullets"—guided missiles, really—to act as specific curative agents. Ehrlich's successful search for the "magic bullet" to cure syphilis came from this idea. Salvarsan was based on a variation of a dye's formula. Connections, connections! In 1858, a chemist by the name of Perkin was trying to exploit the young science of organic chemistry by constructing, synthetically, the molecule of quinine. In doing so, he accidentally created the first coal-tar dye, which started the synthetic dye industry, which started Ehrlich, and eventually resulted in the discovery of new drugs for syphilis and malaria.

37. Syphilis begins with a "boil," the chancre, and usually a rash. In time, the chancre heals and there may be a period of months or even years of clinical quiescence. Meanwhile the treponemes (a.k.a. spirochetes) are beginning to invade the central nervous system.

ing, his eyes dimming with premonition of the convulsion to overwhelm him, crying out to his mother to "give me the sun."[38] Because chemotherapy offered so little hope to those lost to late-stage syphilis, alternate methods of treatment were sought to halt, at least, the progression of the disease (once the nervous system is destroyed by the treponeme, there is no reversal of the severe neurological abnormalities). One such possible alternative was heat treatment.

Not long after Schaudinn discovered the treponeme of syphilis, research began to find techniques to grow the organism in "test tube" culture. During the course of those studies it was observed that *Treponema pallidum* was temperature-sensitive—it was killed when incubated several degrees above the normal body temperature of 98.6°F. Thus, an elevated temperature tolerable to humans would "fry" the microbe. A professor of neurology at the University of Vienna, Julius Wagner von Jauregg, seized upon the possibility of heat therapy for his paretic patients—but how was he to turn up the heat to get a body temperature of 104°F or 105°F for at least several hours? And, of course, there had to be a reliable means of turning off the heat before the patient was roasted. In 1915 there were no

38. Ibsen wrote the play in 1881, twenty-four years before Schaudinn's discovery of the causative organism of syphilis. Nevertheless, at that time the clinical entities of syphilis were well known, as were the modes of transmission. Ibsen would have it that *Ghosts* is about the "sins of the fathers" and Oswald got the infection from his wastrel father. This would mean Oswald's mother was infected during her pregnancy, passed it on to her embryo, and remained herself asymptomatic the rest of her life. It would also mean that Oswald had an asymptomatic infection that became patent in later life. All this is possible, except that Oswald as an infant would have showed some signs of congenital syphilis if the doctors knew what to look for. Equally possible is that Oswald contracted the infection during his two years in Paris of *la vie bohème*—despite his protestations to Mama that he had been a straight arrow during that time.

microwave ovens, or whatever modern medical technology uses, in which you could insert a paretic. But there was one temperature-elevating agent that wonderfully fulfilled all the technical requirements—*the malaria parasite!* There is nothing to equal the malaria parasite, even the "benign" species, in ability to cause fever in the human host—it is the pyrogen *par excellence*.[39] And after the parasite had done its febrile work, quinine was there to produce a relatively rapid parasitological cure and defervescence.

In 1917, von Jauregg began inoculating paretics with blood from patients with benign tertian malaria caused by *Plasmodium vivax*. Several days after receiving the infecting inoculum, the patients began to shiver and burn with their first malarial attack. The fever would break after five or six hours, only to return with clockwork regularity forty-eight hours later. Von Jauregg allowed the cycle to repeat itself three or four times and then gave the curative course of quinine. The effect was remarkable; the downward progression of late-stage syphilis was stopped. Patients didn't necessarily get better, but now they got no worse, and those with minor neurological disorders could return, essentially, to a normal life. Institutions for malaria therapy rapidly proliferated throughout Europe and the technique was taken up in several centers in the United States. In this

39. The mechanism by which the temperature becomes elevated in malaria is not known to any degree of certainty. The current theory holds that it is due to the pyrogenic activity of a soluble factor, interleukin-1, released by macrophages as they periodically attempt to gobble up the parasites released from the ruptured red blood cells. Then again, the mechanism of fever itself is an imperfectly understood phenomenon. The special peculiarity about humans and malaria is how exquisitely sensitive we are, as non or semi-immunes to the parasite. Non-immune humans will have high fevers when there is hardly a parasite to be found in the blood. In contrast, animals with their malarias show much more moderate temperatures even when heavily parasitized.

way, tens of thousands of syphilitics were saved from a sure and agonizing death. For this Wagner von Jauregg was awarded the Nobel Prize in 1927, ten years after he published his first paper describing therapy by malaria.

Malariotherapy was effective, but there were technical difficulties, the chief one being the necessity of having a constant source of parasitized blood. Today, with modern deep freezers, we can store parasitized blood for long periods of time at ultra-low temperatures and thaw it when we want to induce a new infection. However, in von Jauregg's day the malaria infection had, somehow, to be maintained "between paretics" in humans or, where specialized facilities existed, in anopheline mosquito "holding hosts." At that time, there was no way to maintain the parasite in test tube culture as you would a bacterium (even today, *Plasmodium vivax* has defied attempts to maintain it in long-term culture), nor was there an experimental animal—such as a mouse or monkey—in which the human malaria could be "parked" until needed. A solution to the problem came, serendipitously, when in 1932 there was a change in the economics of Calcutta's monkey business.

The Calcutta School of Tropical Medicine had customarily purchased Indian rhesus monkeys for its various experimental needs, but in 1932 those animals became scarce and expensive. To take up the slack, the Calcutta animal dealers glutted the market with Malayan irus monkeys, imported from Singapore, which they sold to the School. The protozoologists at the School who routinely examined the blood of the monkeys in their animal colony found a scanty infection of what appeared to be a new type of malaria parasite in the blood of one of the Malayan monkeys. That animal showed no clinical signs of malaria, reflecting the harmonious balance that nature frequently evolves between a parasite and its natural host. However, when blood from the irus monkey was inoculated into an Indian rhesus, the parasite fulminated in its strange new host and within a few days the rhesus monkey died of overwhelming malaria.

The scientists at the Calcutta School were curious as to whether this new monkey malaria (later named *Plasmodium knowlesi*) could infect humans. Having seen what happened to their rhesus, it must have taken a goodly amount of nerve to attempt to transfer the parasite to humans; but those were the colonial days of insouciant human experimentation. R. Knowles and his colleague, B. M. Das Gupta, inoculated some blood from an infected monkey into a human "volunteer." By blind biological luck the human recipient didn't develop a wild, uncontrollable rhesus-like infection. The parasite caused a high fever for several cycles and then terminated in self-cure without causing any further damage. Moreover, this monkey malaria produced a fever cycle every twenty-four hours rather than the forty-eight hour interval of the benign and malignant tertian malarias of humans.

So here was, potentially, a perfect *Plasmodium* for paretics. It was relatively benign but caused a high fever. Even better, monkey malaria would allow for a shortened, intensive course of therapy because the fever cycles would recur each day rather than every other day of the "human" malarias. It was only a matter of determination and experimental daring to press *Plasmodium knowlesi* into medical service. That first experimental derring-do was carried out in an unexpected place—in Bucharest, Romania.

During the 1920s and 1930s there was an active worldwide network of working malariologists who were aware of each other's research not only through journal publication but also through a lively correspondence between far distant, seemingly unrelated localities. However, in those decades, such remotely separated corners of the world were bound together by the common problem of malaria. Not only did malaria therapy for syphilitics bring the infection to Europe, but Europe itself, from England to Greece, was in the grip of endemic malaria. It was a time when the Rockefeller Foundation was initiating and supporting important, realistic health programs throughout the world, including Europe. Those efforts were being assisted by the League of Nations Malaria Commission.

One of the malaria institutes was in Romania, where malaria was rife and the illness was also used for therapy of paretics. The head of the institute, Professor M. Ciuca, engaged in one of those geographically extended postal exchanges, wrote to John Sinton, working at the Ross Field Experimental Station for Malaria in Karnal, India. Ciuca explained that he would like to try using the newly discovered monkey malaria, *Plasmodium knowlesi*, in his paretic patients. Could Sinton ship an infected monkey to Bucharest? Sinton wrote back that he didn't think an infected monkey would be needed; there wouldn't be any difficulty in shipping infected blood by air, in a viable condition, from Karnal in the Punjab to Bucharest. By air—this in 1935! The weary air traveler of today would be hard put to get from Karnal to Bucharest in a viable condition.

In 1937, Ciuca began inoculating *Plasmodium knowlesi*-infected blood into patients with late-stage syphilis. The monkey malaria fevers were highly successful in halting or attenuating the spirochete's progress. Malariotherapy with monkey and human parasites continued until the mid-1950s, when it was replaced by antibiotic chemotherapy. Thousand of lives were saved by the protozoan parasite battling the spirochete. At England's Horton Hospital alone, between

1922 and the day it closed in 1950, over ten thousand paretics were treated by malariotherapy. There was an additional great benefit: treatment of paretics allowed for highly valuable human studies on the treatment of malaria and the testing of new antimalaria chemotherapeutic agents. In this way, the paretics contributed to the saving of millions of lives in the endemic regions of the tropics.

There was now no doubt that some primate malarias could infect humans. As new species of *Plasmodium* were discovered in monkeys and apes, they were tested in humans for infectivity. By 1966 it had been shown that ten species of *Plasmodium* naturally present in monkeys and apes were capable of infecting man. Most of those transmission trials were done by inoculating blood from the infected primate into the volunteer's vein. The obvious question was whether or not in the real tropical world beyond the laboratory, where man and monkey shared the same territory, the monkeys were giving malaria to their human neighbors. Were anopheline mosquitoes biting monkeys, becoming infected, and then biting humans to transmit the infection?

Certainly, from all the millions of blood slides that had, over the years, been examined for malaria, no one had ever said that they found a malaria parasite that wasn't "human." But that wasn't surprising—it was unlikely that "in nature" there would be anyone to identify such infections. If you are a "local" and get a fever that you think is malaria, you go the health center or malaria office where a blood slide is made and antimalarial pills dispensed if the blood slide is positive. The technicians making and examining the blood films are usually experienced and have a quick and learned eye in picking up the parasite through the microscope's lens. However, they are technicians, not expert malariologists. The technician at the rural health center who might see a "funny-looking" parasite under his (or her) microscope will call it something "human" if only to keep the supervisor off his back. Then, too, some of the differences that distinguish a malaria parasite of monkeys from a malaria parasite of humans can be very subtle, requiring highly expert diagnosis beyond the experience even of most field malariologists.

In the 1960s, malariologists were preoccupied, consumed, by the global campaign to eradicate malaria. In those busy, heady days, little attention was given to the possible zoonotic potential of the simian malarias. The general feeling was that the monkeys were in *their* trees being bitten by *their* mosquitoes and the humans were on *their* turf being bitten by *their* mosquitoes—the two primates neatly separated by the vertically stratified ecosystem compartments. However, in 1960 a mini-outbreak of monkey malaria in humans turned attention to evaluating the zoonotic potential of the Asian monkeys. The infected humans were malariologists, and the outbreak was in Bethesda, Maryland.

Scientists at the National Institutes of Health in Bethesda had been studying the biology of *Plasmodium cynomolgi*, a malaria parasite of Asian monkeys. Under the microscope, *Plasmodium cynomolgi* is an almost exact doppelgänger of the human benign tertian parasite, *Plasmodium vivax*. The inability of *Plasmodium vivax* to infect Asian monkeys indicated that it was a different species, although it was not then known whether or not *Plasmodium cynomolgi* could infect humans. One facet of the research concerned the way the parasite developed in the anopheline mosquito and how it was transmitted by the mosquito from monkey to monkey. To do this, the entomologists had a "domesticated" anopheline that mated and bred readily in caged captivity, and from the Institute's insectary these mosquitoes were supplied for experimental infection. It was all rather a relaxed undertaking and if you visited the National Institutes of Health malaria research laboratory in those days you were likely as not to be bitten by the escapees buzzing about. No one worried about the stray anophelines—even if they were infected, it was only with the monkey *Plasmodium cynomolgi*, which was thought to be of no health risk to the laboratory workers.

On May 5, 1960, one of those laboratory workers, Dr. Don Eyles, called from Memphis, Tennessee, to his friend and colleague at the Institute in Bethesda, Dr. Bob Coatney, to tell him that he had had a fever and headache and that he had examined his blood film and that he had malaria—the malaria, in his opinion, was definitely *Plasmodium cynomolgi*. A few days later several technicians in Coatney's laboratory became febrile and experienced the teeth-chattering rigor of a malaria attack. They too were infected with *Plasmodium cynomolgi*. They had been infected by a mosquito vector in the same manner as if they had been natives of a jungle village in Malaya.

For almost twenty years Coatney had been testing the efficacy of new antimalarials and how they might be used in the prevention and treatment of malaria. The bottom line was always, would it work in the infected human? To get that bottom line, a remarkable program had been established at the Atlanta Federal Prison, in which prisoner volunteers were infected and then treated with experimental antimalarial drugs. It was a program of high American altruism. Malaria was no longer an American health problem, but Americans were offering their bodies and suffering from an uncomfortable, sometimes dangerous, illness in the service of tropical peoples. No deals were cut; the prisoner volunteers were not offered better conditions of their imprisonment or promised any remission of their sentence. Equally astonishing, this program was not only condoned by the government but actually legitimized by an act of Congress (section 4162 of title 18 U.S. Code; Public Law 772 of the 80th Congress). The United States has been the *only* country in the entire world ever to pass a law facilitating human research on malaria.

Coatney, with access to the prisoner volunteers in Atlanta and intrigued by the possible implications of the accidental infections, began infecting humans with *Plasmodium cynomolgi*. These experimental infections quickly revealed that clinically it produced symptoms very much like those of *Plasmodium vivax* malaria—a high fever every forty-eight hours, headache, frequent abdominal pain, and, after a time, enlargement of the spleen. If a Malayan "local" had been infected with *Plasmodium cynomolgi*, there would be no practical way to distinguish it from *Plasmodium vivax*. The zoonotic cloud began to loom on the epidemiological horizon.

In 1965 another event occurred which made the cloud a little darker. It is also a cautionary tale for tourists and business people returned from the tropics and seeking medical care for "weekend" malaria or some other acute travel-related illness. This is the tale of a young American surveyor, a civilian employee of the U.S. Army, who spent five days surveying (at night!) in the Malayan jungle. What a U.S. Army surveyor was doing in a Malayan jungle and what he was surveying at night was never satisfactorily explained. Anyway, after making what must have been a night map of the jungle, he emerges and goes to Kuala Lumpur, where he spends a week being debriefed. Now he's ready to fly home but decides to do this via Bangkok and a few days of intensive R & R. Three days later and well hung-over—there's no town to be on quite like Bangkok—he takes the "flying tunnel" (a Military Air Transport 707 with no windows, seats facing backward, and cheese sandwich / cookie catering) to Travis Air Force Base in California. The plane lands and our surveyor feels bloody awful—chills, sweating, headache, and sore throat. He attributes this to natural phenomena—the indiscretions of Bangkok and the long, steerage-class ride in the flying tunnel—but nevertheless feels sufficiently bad to visit the Base aid station. The busy young physician there summarily diagnoses an upper respiratory infection, gives him some antibiotic pills, and advises him to get the next plane to his home in Silver Spring, Maryland, where he can recuperate from the wear and tear of the Malayan jungle, Patpong Road, and the virus of the URI.

He wakes the next morning, a Saturday, back in Silver Spring, and, if anything, he feels worse than the day before. Worried, he calls his personal physician, a general practitioner. The doctor vaguely remembers from one of the three lectures he had on parasitic diseases in medical school that fever, malaria, and tropical visits go together. He makes a blood film from his patient, and being one of the few doctors still doing some of his own microscopic examinations he sees what he believes to be numerous malaria parasites within the red blood cells. The physician has insufficient knowledge to make an accurate diagnosis, but his reference book tells him that if it is *Plasmodium falciparum* it can be rapidly fatal. The patient needs expert diagnosis, care, and treatment, and because this is a "Government" case, he sends him to the nearby Walter Reed Army Hospital. Reeling from fever, our patient gets to Walter Reed Hospital only to be told that patients are not admitted on the weekend. The Army physician advises him to go cross-town to the National Institutes of Health Clinical Center in Bethesda—they like malaria there and maybe they can treat him. Nauseous, with a 103°F temperature and a blinding headache, our pilgrim makes his way to Bethesda and, finally, to a hospital bed and treatment.

Before beginning antimalarial therapy, the species of parasite—and indeed, whether it was a malarial parasite that was responsible—had to be identified. The laboratory technologist makes a stained blood film and sees malaria parasites that he thinks are *Plasmodium malariae*, a non-fatal quartan infection (fever peeking every seventy-two hours). The attending physician remembers that Bob Coatney at the National Institutes of Health Malaria Unit in Atlanta wants a strain of *Plasmodium malariae* to test its response to antimalarial drugs in prisoner volunteers. So, before chloroquine is given, some blood is drawn from the patient's vein, stored in a fridge, and sent by air to Atlanta on Monday.

On Monday, Coatney inoculates the blood into a prisoner volunteer. A few days later, the volunteer develops the expected fever and Coatney makes a blood smear, expecting to see the red cells infected with *Plasmodium malariae*. To his amazement, Coatney sees the unmistakable morphology of *Plasmodium knowlesi*—a monkey malaria. Clinical confirmation came with the typical daily spiking of fever in the prisoner volunteer. No wonder the surveyor had felt so continuously ill since he left Bangkok. When the story of the infection unfolded, Coatney realized that this was the first confirmed example of zoonotic malaria. A series of improbable events and medical miscalculations had led the parasite into Coatney's expert hands and precipitated the discovery that in nature a monkey malaria *could* infect humans.[40]

Potential to infect was not enough to constitute a health hazard; someone had to return to Malaya to determine whether monkey malaria was actually being transmitted to people living in the *kampongs* (villages) within the jungle. What better man to head the research team assembled by the National Institutes of Health than that wise malariologist with so personal an experience of monkey malaria—Don Eyles? During the 1960s Eyles's team, established at the Institute of Medical Research in Kuala Lumpur, were busy looking at blood films from all sorts of monkeys and gibbons and they were discovering all sorts of new species of primate malaria. Members of Eyles's team inoculated themselves with infected monkey / gibbon blood, and the subsequent chills and shakes amply proved that at least some of the new species, such as the gibbon malaria parasite (later named *Plasmodium eylesi*), were indeed capable of infecting the human animal. These self-inflicted infections made the necessity for field studies even more convincing.

Field work in the Malayan jungle had never been easy. There are trackless expanses, thickly forested mountains, and large rivers whose rapids flow into jungle swamps. This is the ecosystem in which the holidaying Thai silk entrepreneur / wartime OSS agent Jim Thompson went for a morning stroll down a jungle path and disappeared forever without a trace. And the natives were unfriendly (to malaria work-

40. There is a bizarre and cautionary postscript to the story that gives insight to the mixed messages between physicians and patients. The surveyor not only suspected that he had malaria but he had chloroquine with him the entire time from the day he entered the Malayan jungle. He didn't take the chloroquine because he remembered a doctor telling him never to take a drug of any sort without a physician's explicit advice.

ers). Malaya was then in the throes of its national malaria eradication campaign and the *kampong* dwellers were tired of having their fingers stuck repeatedly in service of the promise of a malaria-free life that never seemed to arrive. What was even worse than a sore finger was the strange phenomenon of their roofs collapsing within a month of their houses being sprayed with DDT.

Malaria they knew, but falling roofs were something else again, and the sophisticated biological explanations offered by the malaria workers didn't put back the cover on the house. What was happening was that the roofs were made of *attap* (palm fronds) and there was an *attap*-devouring caterpillar that dwelt in the roof. Under normal conditions a parasitic wasp preyed on the caterpillars and kept the pests at low, non-destructive numbers. Unfortunately, the wasp was highly sensitive to DDT while the caterpillar was resistant. The malaria workers sprayed the houses, the wasps died, the caterpillars proliferated . . . and the roofs came tumbling down. Malaria workers were told that they weren't welcome in the *kampongs* and forcibly ejected if the polite dismissal wasn't heeded.

The problems of epidemiological diplomacy were more than equaled by the experimental problems that would be needed to determine whether monkey-to-man transmission of malaria was taking place in its natural setting. It was one thing to conduct tidy transmission studies under laboratory and hospital conditions, but quite another matter to enter a village, bleed the populace, and prove that some of the infections were of monkey origin. There was, at that time, no genetic-marker, biotechnical methods to distinguish look-alike species from one another. The only, and still the most convincing, method was to take blood from the villagers and inoculate "clean" monkeys (which would have to be rhesus imported from an area of India where primate malaria doesn't exist). Any consequent infections in the recipient animals would be proof of zoonotic monkey malaria (since, you may recall, truly human malaria parasites are not infective to Asian monkeys).

This Herculean experiment was actually carried out by a group of the National Institutes of Health team, led by MacWilson Warren, an affable young man. Mac gained the trust of the villagers by his genuine concern and affection for all things Malay, and by passing the trial by fire—the ability to consume, with relish, the mega-chili curries of the Malayan village cuisine. The monkeys were obtained from India and housed at the Kuala Lumpur Institute. Then the team drew venous blood from twelve hundred jungle *kampong* dwellers (an extraordinary feat in itself, considering the great reluctance of the rural Malay to have blood drawn from the vein). The blood samples were iced, rushed back to Kuala Lumpur, and inoculated into the rhesus. No monkey ever became infected. Humans were not infected with any monkey malaria. The research team came to the conclusion that in Malaya, at least, monkeys were not giving malaria to man and that antimalaria campaigns did not have to incorporate strategies to control transmission from lower primates.

That was in the 1960s; it is not known whether the assumption that monkeys are not a threat to human health in respect of malaria holds true today. Since that time the Malays have cut down their forests with the same pillaging pace as in other parts of the tropics. The ecosystem of the rain forest is being destroyed, and with it the arboreal homes of the resident monkeys. But some of these monkeys are true Asians—survivors—intelligent, tough creatures who can adapt to new conditions. They can even become city dwellers, where they form bands of urchin thief-beggars protecting their territory like their street-wise human counterparts.

Ecological change has brought monkey and man closer together. Forest and field are now more blurred and at this new ecotone the monkey's domain blends with that of the human. In Africa, it is at such altered ecotones, created by the destruction of the rain forest, that newly semi-domesticated monkeys and mosquitoes have departed from the sylvatic life to bring another vector-borne disease to their human neighbors—yellow fever. Whether or not recent ecological changes have brought a similar change in monkey-mosquito-man transactions as regards zoonotic malaria may never be known. There are no longer the resources, interest, expertise, or will to carry out the kind of expensive / extensive studies conducted by Eyles's group in Malaya. The emphasis of scientists studying malaria today has shifted to the molecular level. It is still the M&Ms, but now, too often, that has been translated as "malaria, the money-making molecule."

Chapter 10

Mal'aria –The Bad News Airs

DEAD MEN tell no tales, and when malaria has caused their death the microscopic killers vanish without a trace. Malaria leaves no hallmark on the bones, no "tree rings" from which paleoanthropologists can diagnose malaria thousands of years later, from skeletal remains. Thus there is no way, beyond imperfect speculation, that we can know when malaria first came to our ancestors or how remote those ancestors were. The human cradle was in Africa—we are all Afro-Americans (and Afro-Europeans- and Afro-Asians). Our origins are in Africa, beginning with the bifurcation into the anthropoid ape and anthropoid hominid branches some 4.5 million years ago.

Then, about 2 million years ago, came Lucy of the Olduvai Gorge—a pert 3 feet 5 inches, small-brained but big-headed, an amalgam of ape and human physiques. There followed a succession of man-apes until 1.5 million years ago the Lucys of Africa were replaced by *Homo erectus*, a bigger-brained creature, at 5 feet 5 inches, but still barrel-chested like an ape. But *Homo erectus* acquired a behavioral trait that was to set him apart from his anthropoid and hominid ancestors. It was a very human trait that a long time later was to establish the tourist industry. *Homo erec-*

tus lost the territorial imperative of the animal and began to wander at will—like a human.

The Red Sea need not have parted for the migrating bands of *Homo erectus*. One million years ago when *Erectus* made that crucial journey out of Africa to Asia, there was a land bridge between the two continents. From East Asia, *Erectus* wandered throughout the continent (he is the Peking Man of early archeological fame). In Asia, *Erectus* man may have first become malarious. There are so few monkey malarias in Africa, and those that are present are so unlike the parasites of humans, that logic would dictate that as humans were beginning to "grow up" in Africa, malaria was not there to burden their evolutionary progress. In contrast, as the preceding chapter has noted, Asia, particularly Southeast Asia, is the "home and birthplace" of numerous species of primate malaria, some of which are morphologically analogous to parasites of present-day humans and are, in fact, capable of infecting humans. Two of those monkey malaria parasites became "humanized"—*Plasmodium cynomolgi* became *Plasmodium vivax* (the benign tertian malaria), and *Plasmodium inui* became *Plasmodium malariae*, the quartan malaria.[41]

It is not known whether it was in Asia or Africa that man became truly human, evolving first to *Homo sapiens neanderthalis* and then, if Dr. Becky Cann and her DNA detectives of the University of Hawaii are right, a single woman

41. These are morbid but not mortal parasites—they rarely kill; even non-immunes recover. During the course of evolution, humans acquired natural defensive blood traits, such as sickle cell trait, Duffy factor, and glucose 6-phosphate dehydrogenase deficiency that made them even more resistant to the effects of infection—see chapter 5, "The Bean, the Gene, and Malaria," in my earlier book, *New Guinea Tapeworms and Jewish Grandmothers*. Thus, the infection of *Homo erectus* with this new disease—malaria—would not have been a serious threat to the species.

mutated 200,000 years ago to found the entire human race of *Homo sapiens*. So maybe male chauvinism mistranslated the Old Testament and it was really Eve who came first and it was *she* who gave her rib (in its DNA form) to make an Adam. The men, Neanderthal and Sapiens, spread from Asia to Eurasia and then to Europe, presumably taking their malaria parasites with them; parasites to be perpetuated by the awaiting anopheline mosquitoes indigenous in areas as far north as southern Siberia and Britain.

Elsewhere in this book I note that from colonial times until the 1940s, malaria was *the* American disease. One of the first military expenditures of the Continental Congress was for $300 to buy quinine to protect General Washington's troops. During the Civil War one half of the white troops and four fifths of the black soldiers of the Union Army got malaria annually. The tropical Americas remain highly malarious. When did malaria first come to our Western Hemisphere shores? Distinguished malariologists have long debated whether or not the Western world's peoples or its primates had malaria before Columbus opened the flood of immigration from the white and black Old World. The majority of opinion has it that there was no pre-Columbian malaria in man or monkey; the parasite probably was "frozen out" in the Asian migrants as they crossed into Alaska and slowly, over 20,000 or more years, made their way south to Tierra del Fuego. There are no records of malaria in the "medical books" of the Mayans, Olmecs, or Aztecs, and it is doubtful whether those great civilizations could have arisen and flourished in a malarious setting. When, in 1519, Cortés rested his troops on the isthmus of Panama before beginning his conquest of Mexico, he makes no mention in reports to his king, Charles V, of malaria as a health problem. Within two generations after the Conquest, from the sixteenth century onward, European settlers (remember, even the Thames estuary was highly malarious until the turn of the twentieth century) and their "peculiar institution"—slavery—would have repeatedly imported their malaria parasites into the New World and the awaiting anophelines of the Americas.

The fitting together of our theory on the origins of human malarias has, so far, been based upon logical assumptions from knowledge of human evolution and migrations, and the biology of primate malaria parasites. The plot, the line of reason, falls apart when we come to the "real" malaria, the "Mother of fevers" as it was known to the ancient Chinese—*Plasmodium falciparum*, the malignant tertian malaria parasite. It is unique and there are no other parasites of bird or beast that bear a resemblance to it, except a malaria of the gorilla (*Plasmodium reichenowi*, which is incapable of causing infection in humans). *Plasmodium falciparum* has a DNA homology that suggests it to be closer to the malaria parasites of birds than those of monkeys.

How then did the first humans who returned to Africa manage to survive and, finally, flourish, despite exposure to lethal malignant tertian (falciparum) malaria? Their preservation may well have been due to the as yet undisturbed high forest ecosystem in which they lived and the relatively small size of their pre-agricultural, nomadic, hunter-gatherer bands. The anopheline mosquito that was to become the chief vector of malaria in Africa, *Anopheles gambiae*, was present, but not in menacing numbers. *Anopheles gambiae* selects small, sunlit collections of water to lay its eggs. The intact high forest provides few such breeding sites, and in the dark, dank jungle there are few malarious mosquitoes. The hunter-gatherer life style was also antimalarial. The group numbers were sufficiently small that there would be few malaria carriers to recirculate the infection through the mosquito. Then, too, the hunter-gatherer is nomadic and doesn't stay in one place long enough to infect the local mosquitoes. Early man in Africa may have had malaria, but if our scenario is true, there would have been few cases and still fewer deaths—so long as they persisted as hunter-gatherers and so long as they protected the integrity of their nurturing high forest. Two thousand years ago Africans began to forsake that way of existence and they began to disassemble their environment. And they created the conditions of their destruction by life-threatening malaria.

That alteration of social and ecological structure was brought about by another human migration from Asia. The Malays, by a series of incredible voyages across 4,000 miles of open ocean in their double-hulled sailing canoes, colonized Madagascar. They brought with them new root and tree crops—yams, taros, bananas, coconuts. This in turn introduced the sense of settlement; agriculture replaced foraging. These easy-to-grow crops soon reached the African mainland, where they were enthusiastically adopted by the blacks of the forest. Now began the destruction of Africa's tropical forests as the trees came down to yield to plots of taro and yam. This created wet-islands within the forests ideal for the breeding of the malaria-transmitting *Anopheles gambiae*. These female mosquitoes now had a domesticated, steady food supply of blood—the settled human agriculturists. It may well be that malaria would have destroyed these first African agricultural pioneers if it were not for another migration, and another infusion of a protective gene.

In ancient times, a land bridge across the Red Sea bound Asia to Africa. From their home in India, aboriginal Veddoids came into Africa carrying with them the gene that coded for the manufacture of an abnormal form of hemoglobin—the "working" constituent of red blood cells. This was the sickle cell gene. In its "purest" form, the double dose of the gene inherited from both parents causes eventual death from progressive sickle cell anemia. However, in its mixed, diluted form in an individual who inherits a normal hemoglobin gene from one parent and a sickle cell hemoglobin gene from the other parent, it is protective against the full pathogenic force of falciparum malaria. This is the "sickle cell trait." Children with sickle cell trait are as susceptible to *infection* with *Plasmodium falciparum* as children with normal hemoglobin, but the parasite does not flourish in the sickle cell trait erythrocyte; the attack is attenuated, and the child survives to eventually become an adult protected by an acquired immunity.

It is difficult, if not immoral, to speak dispassionately of "balanced deaths," but this is what the sickle cell gene confers to a poor agricultural community. It allows the survival of the group; the cost of that survival being a high death rate from malignant malaria in children with normal red cell hemoglobin, and death from sickle cell anemia of the children with the double dose of the gene. The sickle cell gene not only permitted community survival, but by ensuring a continuous supply of "carriers" of the parasite also established the skein of malaria transmission in Africa that remains unbroken. And it is here that we end our speculative scenario of circumstantial evidence and turn to "harder" facts of malaria's history.

Those harder facts came with the moving finger that began inscribing the written word. From the time humans began writing, 6000–5500 B.C., they left a record of their medical complaints. They kvetched on tablet, papyrus, and parchment. The first civilization, the Sumerian, rose in the lush valley between the Tigris and Euphrates rivers. Lush and malarious—the cuneiform medical writings of the Sumerians repeatedly describe the typical fevers of malaria. The neighboring Jordan Valley was also intensely malarious. In biblical times, the pillaging Assyrians came upon the Jordan "like a wolf on the fold," but in Old Testament plots, like movie morality of the pre-1960s, the bad guys couldn't win and the Angel of Death smote the Assyrian hordes. Modern malariologists now reason that the Dark Angel assumed the guise of the malignant tertian malarial parasite.

Inexplicably, neither the Old Testament nor the Talmud, holy script and commentaries that are so rich in the medical history of biblical times, make mention of malaria as an affliction of the Hebrews either in Israel or the lands of their bondage. The Jews were taken into captivity through the malarious Jordan Valley, through the malarious Euphrates Valley into malarious Babylon, but not a written word of complaint to Yahweh that they suffered of the ague. Maybe they just didn't want to talk about a foreign medical problem. A stiff-necked people, the Jews in their Diaspora have always refused to assign Hebrew names to the natural attributes of their homeland-in-exile. A snake in the Sinai got a Hebrew name; a snake in Poland did not. Perhaps this taxonomic negativism extended to their diseases. Or, perhaps, malaria just wasn't a Jewish disease.

We also know that about the same time as the Sumerians were shivering and sweating from their malaria fevers, the Chinese too were similarly afflicted. The Chinese medical classic of 2700 B.C., the *Nei Ching*, gives an accurate description of tertian and quartan malaria fevers and notes the enlargement of the spleen that follows the attack. The *Nei Ching* maintained that these malarial fevers were caused by the Yin and Yang being out of whack. Curatives made from medicinal plants were prescribed.

The Vedic writings of 1600 B.C. with their many references to deadly fevers indicate that India too was already in the grip of malaria. However, at that time Europe was still malaria-free, and it was to be another one thousand years before the disease invaded Greece and began the engulfment of the continent. By 600 B.C. the great cities of Greece were conducting their conflicts and commerce with the malarious countries of Mediterranean Asia and Africa. This was the period when the glory that was Greece was truly glorious. Thus, it was a time when malarious people—traders, slaves, soldiers—would have repeatedly come to Greece to introduce the parasite to the awaiting Hellenic anophelines. By the fourth century B.C., malaria was a major health problem and medical historians speculate that it contributed to the demise of that civilization. The great doctor of that time, Hippocrates, gave a highly accurate account of the intermittent fevers which he recognized as a distinct disease entity. Hippocrates also made the broad association between the disease and the environment. He was the first epidemiologist and discerned that the intermittent fever (malaria) cases were clustered in swampy areas.[42]

Hippocrates, in keeping with the current "scientific" thinking of the day, believed the intermittent fevers were due to the body's humors (blood, phlegm, black bile, yellow bile), which were out of sync. Hippocrates also saw the effects of chronic malaria, which he believed was caused by

42. In Europe, the anopheline mosquito carriers of malaria breed mainly in marshy wetlands. Malaria in Europe was a "swamp fever." However, in Asia, Africa, and other endemic regions, other anopheline vectors breed in a large, diverse variety of waters and place, ranging from small puddles to streams to lakes and inundated rice paddies.

drinking stagnant marsh waters (which, in turn, upset the humors). If you visited a malaria-stricken village today, his words, written 2500 years ago, would reverberate in your mind as you examined those with "large, stiff spleens, and hard, thin hot stomachs while their shoulders, collarbones and faces are emaciated." But the genius of Hippocrates went one step further: he shrewdly guessed that there was some outside agent that was causing the disordered imbalance of the humors. Was there a miasma, a dank effluvium rising from the marshes, that was the primary causative factor of intermittent fevers?

By 200 B.C. the mantle of power and civilization—and malaria—had passed from Athens to Rome. Three hundred years earlier, the scattered villages clinging to the hills rising from the left bank of the Tiber began to coalesce into what would become the Eternal City. Under the hills lay the Pontine marshes (later, the Campagna di Roma). Beyond, from the Apennines to the Tyrrhenian Sea, lay the plains of the Latium. From about 200 B.C. until the early 1930s when Mussolini drained the marsh, the Campagna was one of the most malarious areas in the world. There was such a close association between malaria and Rome that, in a sense, the Romans viewed the disease in a proprietary fashion—it was their "Roman fever." Even the term "malaria" is of Roman origin. The disease was not known by its present name until the mid-eighteenth century (before then it was known variously as the ague, intermittent fever, swamp fever, Roman fever, death fever, etc.). There is some argument over who was first to use the word *malaria*—"bad" or "evil" air; a name deriving from the miasma theory of causation. The first reference was probably made by the English writer, Horace Walpole, who wrote from Rome in 1740 of "A horrid thing called mal'aria that comes to Rome every summer and kills one." In their medical books published in Rome, P. F. Jacquier in 1743 and Francisco Torti in 1753 write of *mal'aria.* Later, during the late nineteenth and early twentieth centuries, when proud, contentious scientist-doctors from Britain, France, and other countries wanted to study malaria, ultimately all roads led to Rome.

Chapter 11

Malaria: From the Miasma to the Mosquito

I CARRIED OUT a routine laboratory procedure this morning. A blood film was sent to me from the hospital. It was from a forty-two-year-old woman who, two weeks ago, had been on holiday exploring rural Thailand. Late last night she came to the hospital emergency room complaining of feeling feverish, shiverish, and with a splitting headache. The emergency room physician, a young graduate of our medical school, recalled the words of his old professor of parasitology ("Always get a travel history," and "Think of malaria in travel cases of unexplained fever," and "Malaria is the great mimic, it can present symptoms like almost anything else but a broken leg") and had the woman admitted to hospital with instructions that a blood film be sent to his old professor of parasitology to confirm the diagnosis. The blood film was fixed in methanol, stained in Giemsa's stain for an hour, and then dried with that indispensable laboratory instrument, a hair dryer. It was set on the stage of the Leitz binocular research microscope, a drop of oil placed on the film, the high-powered, oil-immersion objective lens carefully lowered onto the slide, and the powerful illumination switched on to shine through its optical condensing lens system.

I peered through the flat-field eyepiece lenses and saw

the straw-colored red blood cells and the white blood cells with their vividly stained blue cytoplasm and red nucleus. Then, as I twiddled the knobs of the finely engineered mechanical stage to move the slide to a new field of view, I spotted within some red cells the unmistakable forms of *Plasmodium falciparum*, like minute signet rings, with a small ruby nucleus embedded in a wisp of blue-staining circlet. There were a few banana-shaped forms, blue with a red nucleus. These were the gametocytes, the sexual forms that would develop further within the anopheline mosquito vector. I also knew that, unseen, there would be developing "sporulating"-stage forms—the schizonts—confined in the deep recesses of the small vessels of certain organs. The lady had malaria, and I called in my diagnosis with some informal advice on treatment schedule.

The procedure was routine, almost mindless; any malariologist or well-trained technician could pick out the parasites, distinguish the different growth stages of the developmental cycle, and from their "anatomical" landmarks diagnose which of the four species of human malaria parasites was present. I made the microscopical diagnosis and passed on to the next order of business without further thought. However, the other enterprise of the day was the beginning of this chapter on the historical process by which the long, confused, often contentious years of scientific research led to the discovery of the true nature of malaria. How bewildered those early researchers must have been by the multitude of shapes and forms of these protozoan parasites that were still so imperfectly understood. And they began the laborious work of sorting out the parasites and their life cycles with the instruments of their day—microscopes so feeble in magnification and clarity that I would not give one to my six-year-old granddaughter as a Christmas present.

Even during the long, almost 5000-year "pre-microbial night," there had been explanations. All malaria-affected cultures have figured out, with absolute confidence, the cause of the disease. The Chinese were certain that the paroxysmal fever—malaria—was caused by the disharmony between the Yin and Yang. Hippocrates and his followers were certain that the swamp miasmas and the stagnant waters caused an imbalance of the humors to produce malaria. The false miasma of malaria persists as a cultural belief, often to the despair of malariologists. How are you going to convince a population to take antimalarial pills, or sleep under a bed net, or have their houses sprayed, when they have the certain conviction that the cause of malaria is a swamp poison, drinking stagnant water or water poisoned by the urine of a green monkey, or eating a green banana? And, mind you, to these people their theory is as scientific as the microbial etiology is to you. They would be derisive if you asked them whether they thought malaria was caused by the evil intent of demons or vengeful gods. Such entrenched beliefs can seriously affect the course of a health campaign. For example, in 1966 the Philippine national malaria eradication program all but collapsed because the villagers refused to give finger-stick blood for microscopical diagnosis or have their houses sprayed with insecticide. Why should they be subjected to this nonsense when everyone knew that malaria was caused by a combination of polluted water and fatigue?[43]

A fascinating assortment of etiological theories continued to be offered during those pre-Pasteur centuries. About 50 B.C. a concerned Roman husband, Marcus Tentius Varro, wrote a little how to be fit and healthy book for his beloved

43. The fact that the Philippine malaria program had switched, several years earlier, from DDT to the more toxic dieldrin insecticide which killed the domestic animals of the villagers didn't help matters, either.

wife Fundavia. He warns Fundavia that she should stay away from the low-lying swampy areas, for here breed minute unseen animals that infest the air and drinking water. These were the animals that caused the fever of malaria when they entered the human body. The details, of course, were queer, but the public health advice was as good as you can get today, and Mr. Varro was prescient in his opinion that malaria was a disease caused by the unseen, by the pathogenic microorganisms.

Probably the last pre-microbial theory of malaria was that of Carolus Linnaeus, the Swedish botanist-physician to whom we pay homage each time we call any living thing by its proper scientific name or classify it according to its place within the plant or animal kingdom. Linnaeus was an intellect of the Enlightenment; natural history and medicine were inseparable, and it is difficult to imagine the physician of today, specialized and mono-minded, being as professionally bifurcated as Linnaeus. At the University of Uppsala he was, simultaneously, professor of medicine and professor of botany. Linnaeus was a classifier, a taxonomist, of both living things and disease. He devised the binomial nomenclature of biology (such as the genus-species designation of the modern human, *Homo sapiens*). Although essentially a taxonomist, in 1735 he wrote his Doctor of Medicine dissertation on malaria for the University of Harderwijk. Morbid anatomists and other early physicians since the time of Hippocrates had observed that the spleen, liver, and brain of those who died of acute "swamp fever" had a grayish-black color.[44] Hippocrates had reasoned that color was caused by an overabundance of black bile, the bad humor

44. Within the red blood cell, the *Plasmodium* actively ingests and digests hemoglobin. The inert "feces" is the metabolic product hematin, which is deposited in the organs where infected red blood cells tend to collect (the spleen, brain, liver) and gives them their grayish cast.

of malaria. In his doctoral dissertation, Linnaeus argued differently—the humors were beginning to lose their pervasive grip on medical philosophy by the eighteenth century. Linnaeus looked at the blackish organs and was impressed by their clayey appearance. His imagination soaring, like any good graduate student, he went on to hypothesize that when a person drank water containing suspended particles of clay, those particles came to lodge in the small blood vessels of the organs. The occlusion and irritation then gave rise to the symptoms of malaria.

However, even at the moment Linnaeus was proposing that malaria was inanimate—nothing more than common clay—the artisanship and science of optics was developing to bring the unseen world of the minutiae, imagined by men like Varro, into startling, teeming reality.

In 1674, the lens grinder of Delft, Anton van Leeuwenhoek, brought the microbial world into focus. This marvelous Dutchman had an insatiable curiosity for the minuscule world he viewed under his home-made microscopes. In a long series of communications to the Royal Society of London he described the "animalcules" that he saw in all sorts of collections—in rainwater, pepper infusions, his own decayed tooth, frog feces, and, unwittingly, the first protozoan pathogen, *Giardia lamblia,* in his diarrheic stool. However, it was to take another two hundred years before Pasteur would reveal that some of the "animalcules" were pathogens—the sick-making yeast that turned good wine bad, the protozoan killers of silk moths, and, at last, the bacterial microbes of animals and humans. By 1870 Pasteur, followed by others, notably Robert Koch and his school in Germany, had firmly established the revolutionary new principle of specific microbes (chiefly bacteria) as causatory agents of specific diseases. It was a time of great discoveries that prompted the general notion that a specific bacterium was the cause of each of the many, if not all, diseases. One had only to be diligent enough in isolating and identifying the microbe.

Since in the scientific *Zeitgeist* of the 1870s all diseases were bacterial, it was not unexpected that someone should "find" the bacillus of malaria. In 1879 Edwin Klebs and Corrado Tomasi-Crudeli, working in Italy, approached the problem from the established "fact" that malaria was a disease of the swamps and therefore the bacterium must be either in the waters of the swamp or suspended in the dank effluvium. They took some water from the Pontine marshes and injected it into rabbits. The rabbits got sick, became feverish, and had enlarged spleens. A bacterium was isolated from the dying rabbits. Klebs and Tomasi-Crudeli were convinced that it was the bacterium of malaria and named it *Bacillus malariae*. The scientific community applauded and welcomed the discovery even though no one could repeat the findings. The long search for the cause of malaria was ended. It was a bacterium. Wishing had made it so.[45]

A year later, in 1880, the first true sighting of the malaria parasite was made by a French Army physician in Algeria. That sighting was only the bare beginning; it was to take another seventy years of research and confusion and sometimes chaos before the life cycle of the malaria parasites was fully elucidated. The historical road to that elucidation is fascinating, but it may be prudent to pause in our story here and give a synoptic account of the malaria parasite's

45. Even today there are examples of "wishing making it so" from imperfect evidence. For years it was accepted as the oncological gospel that the herpes virus caused cervical cancer. Only recently has it been proven that a completely different virus, the papilloma virus, is the true cause. And there are a few dissidents who adamantly maintain that the HIV virus is not the cause of AIDS and that the real culprit remains undiscovered.

life history, a kind of libretto to the dramatic opera *Mal'aria,* or *Mlaria,* as it is called on the vanity license plate of one present researcher. For those of you who would never prematurely turn to the last page of a whodunit, the biological program notes can be ignored and you can turn directly to the adventures of Dr. Laveran in Bone.

The essentials of the malaria parasite's life history are illustrated in the accompanying cartoon (figures 1–6). Humans (except, as noted, for the very rare natural infection with a monkey malaria) are host to four species of malaria parasite: *Plasmodium falciparum, Plasmodium vivax, Plasmodium ovale,*[46] and *Plasmodium malariae.* While there are considerable major differences between the species in their pathogenicity (*Plasmodium falciparum* is the only one that is highly virulent and potentially lethal) and epidemiology, as well as subtle, but important, differences in appearance, development, and host-parasite relationships, all four species share a common, basic life cycle.

The infection in humans begins when an infected female anopheline mosquito (only the lady mosquito partakes of blood; the male, gentle fellow that he is, flies about in a lifelong pursuit of sex and nectar) injects into the bloodstream, during the act of feeding, threadlike malaria parasites (the *sporozoites*) that have been stored in her salivary glands. Thousands of sporozoites are usually injected (figure 1) and they are carried in the bloodstream to the liver, where they leave the circulatory system and each sporozoite penetrates a "building block" cell of the liver tissue. Within the liver cell the sporozoite rounds up and transforms into a "spore." For about two weeks this spore repli-

46. *Plasmodium ovale* resembles *Plasmodium vivax* in life cycle and clinical effects. It is a relatively rare malaria parasite that most frequently occurs in endemic foci of West Africa.

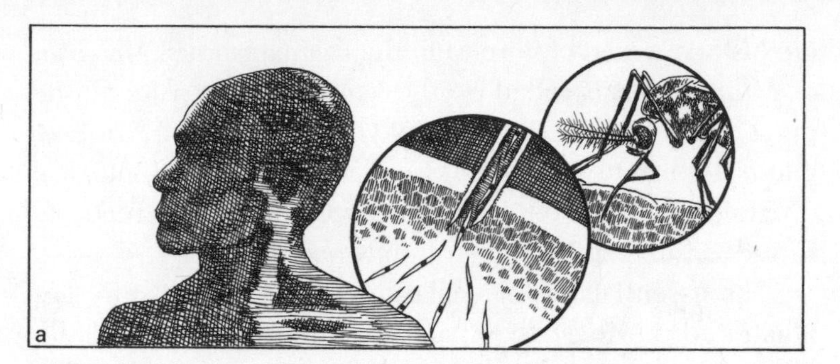

FIGURE 1

cates repeatedly (a process known as *schizogony*), until there are many thousands "spores" *(merozoites)* within a cystlike structure, the host liver cell having been destroyed by the proliferating parasites (figure 2).

Those two weeks are a clinically quiescent period for the person within whom the seeds of malaria are undergoing repeated division. There is no fever, no sign of the illness that is to descend so swiftly. The first clinical attack—the intense rigor and sweating with high fever—develops when the cyst bursts to release the myriad of "spores" (mer-

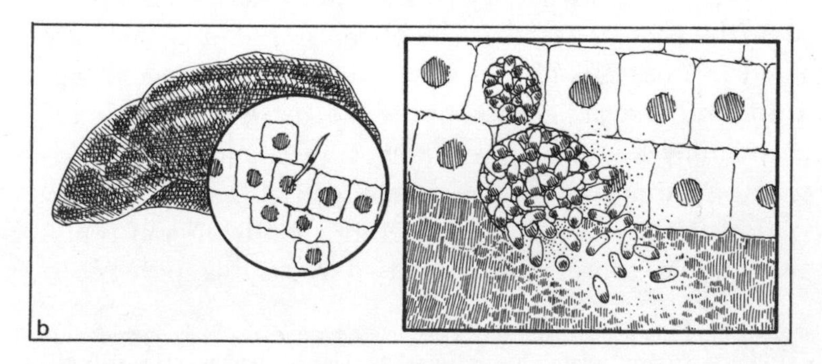

FIGURE 2

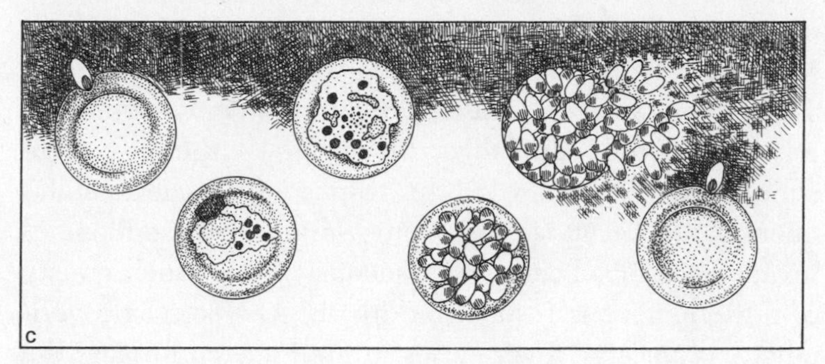

FIGURE 3

ozoites) into the bloodstream (figure 3). Each merozoite now attaches to the surface of a red blood cell and then enters it. Inside the red blood cell the young parasite appears as a minute circlet with a nucleus-dot (the ring stage). The parasite feeds avidly by engulfing, in amoeba-like fashion, the red cell's hemoglobin. The parasite grows, the "body" cytoplasm increasing until it fills more than half the red cell (the *trophozoite* stage).

The next developmental event is the asexual "shattering" of the nucleus *(schizogony)* into eight to twenty-four discrete bits (the number depending of the species of malaria parasite) within the cytoplasmic matrix. There then occurs a complex reorganization in which the cytoplasm coalesces around each nuclear bit to form a "spore," that is, a merozoite. The demolished red cell bursts; the merozoites are released into the bloodstream to attack and invade new red blood cells.

This process is repeated over and over again with more and more red cells becoming parasitized until natural or acquired immunity, or antimalarial chemotherapy, or death (in the case of untreated *Plasmodium falciparum* infections in non-immunes) brings the repetitive process to an end. Moreover, there is a marvelous synchrony in development. The growing malaria parasites are a *corps de ballet* moving together in their growth cycle; all are at the ring stage simultaneously, all are at the trophozoite stage simultaneously, all burst, as merozoites, from their millions of invaded red blood cells simultaneously. This synchronicity of development is responsible for the characteristic periodicity of malarial fever cycles in the infected human; the forty-eight hours between fever peaks for *Plasmodium falciparum*, *Plasmodium vivax*, and *Plasmodium ovale*, and seventy-two hours between peaks for *Plasmodium malariae* malaria.

After several asexual generational cycles, some of the merozoites become committed to a completely different form of development. They become the male and female sexual stages (the *gametocytes*). What underlying mechanism drives some merozoites to become sexual and what genetic reorganization must take place to direct them to become so different in form and function remains an intriguing mystery. The "sexual" merozoite invades a red blood cell to become a ring stage, like all other ring-stage parasites. But instead of the nucleus dividing, it becomes greatly enlarged and the cytoplasm about it increases to form a discrete body filling, or almost filling, the host red blood cell (figure 4). These gametocytes circulate in the bloodstream undergoing no further change until taken into the stomach of the feeding female anopheline mosquito.

Inside the mosquito's stomach, all the asexual-form parasites die as their host red blood cells are digested. However, the gametocytes are now in *their* milieu as they become naked of the host red blood cell membrane. The female gametocyte *(gamete)* remains quiescent, although the care-

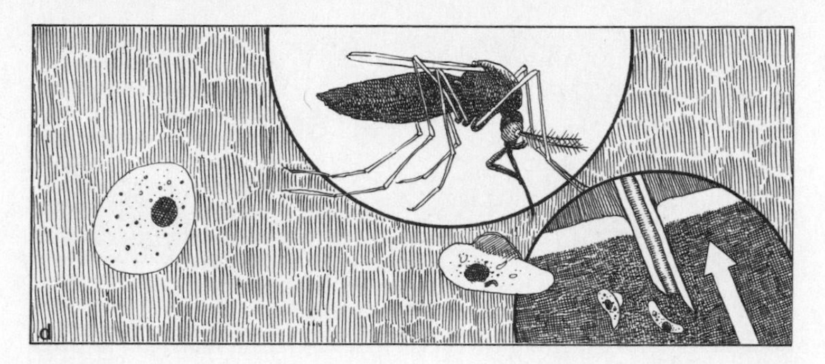

FIGURE 4

ful eye and special, powerful microscopical optics reveal her quivering, scintillating cytoplasm. The male gametocyte undergoes a spectacular development: his nucleus divides repeatedly, the cytoplasm reorganizes, and from the surface of the gametocyte there emerge numerous lashing filaments. These are the "sperm," the male gametes, and the process of development from the gametocyte is called *ex-flagellation*. These sperm leave the "father body," swim to a female "egg"-gamete, penetrate, and fertilize her (figure 5).

The fertilized egg elongates to become a creepy mobile form that older malariologists gave the whimsical designation of the "traveling vermicule." This vermicule makes its way through the insect's stomach wall to come to adherent rest on the exterior surface. There it rounds up to become a minute, pearly-looking cyst (the *oocyst*). Within the oocyst still another intense nuclear-cytoplasmic reorganization takes place over a period of fourteen to twenty-one days (the time depending upon the temperature, species of malaria parasite, and species of mosquito), to end in the formation of several thousand of the infective, threadlike forms—the

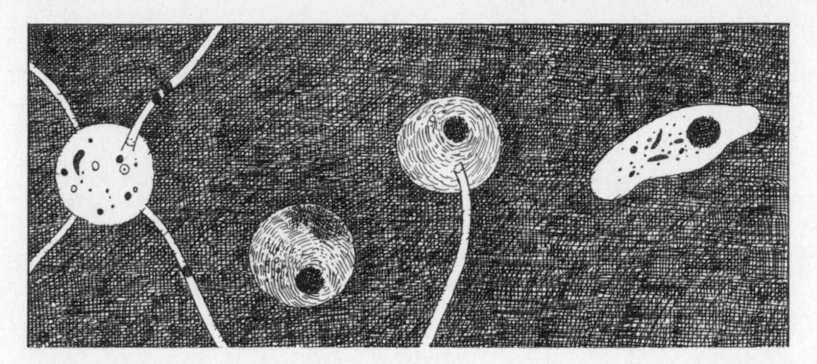

FIGURE 5

sporozoites. The mature oocyst now bursts, releasing the sporozoites into the mosquito's body cavity. The salivary glands, whose purpose it is to inject anticoagulant saliva so that the blood pool will not clot and choke the insect, extend into the body cavity. The sporozoites invade the salivary glands and the anopheline mosquito is now "loaded," the sporozoites being injected with the saliva when the mosquito next feeds on a human host (figure 6). That's how you

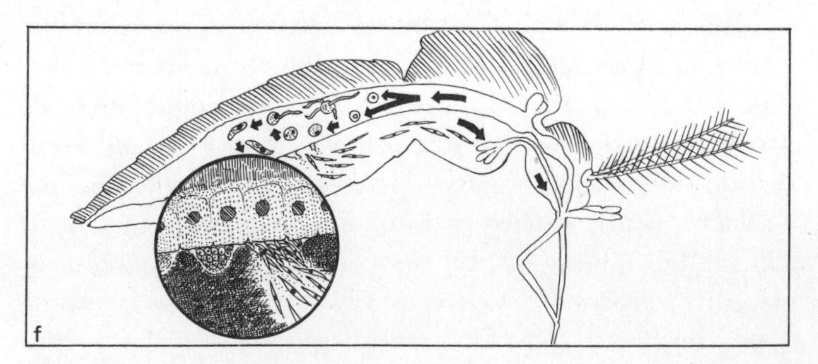

FIGURE 6

get malaria.[47] Now, let us return to the historical stream of the narrative.

Armies have a great need for physicians. In the wake of battle the military doctor has an instant, large clientele who require limbs lopped off, arrow and bullets extracted, wounds dressed, et cetera. Between engagements, the army doctor also has, generally, a busier time than the family practitioner (although I am sure that the Roman legionnaire had as low an opinion of military medicine as the World War II GI of my era—aspirin for any complaint above the waist and calamine lotion for anything below). Armies have been particularly beset by infectious diseases which have been vexatious to treat and control. The closeness of barracks life makes for the rapid dissemination of airborne infections. A soldier's life is terribly hard and traditional relief comes from recreational sex. Another problem for the military physician. Farwell, in his book *Armies of the Raj* (1989), notes that in the nineteenth century, approximately 50 percent of the British troops stationed in India were hospitalized for a venereal disease *each year*. Finally, troops stationed in foreign parts were / are at risk to acquiring foreign diseases. The military is, therefore, greatly concerned with the field which we in the West call "tropical medicine." A good deal of important research on the nature, treatment, and control of the diseases of the tropics has emanated from the military establishments of many nations. The causative agent of many important and perplexing diseases of native

47. Mosquito transmission is the "normal" way of contracting malaria. Malaria can also be acquired by being the recipient of infect blood received for medical transfusion or even in the exchange of the shared needles of drug addicts. And malaria can be acquired congenitally when the parasites cross the placenta from an infected mother to her fetus. This is a rare occurrence, however, even in highly endemic regions.

populations has been elucidated from the infected soldiers of occupying armies by the foreigner-physicians of those armies. We noted earlier in this book that the telltale sign that first revealed *Leishmania donovani* as the cause of kala azar was such a sacrificial soldier. And so it was with malaria.

Military families tend to form a caste, passing the profession of arms from father to son, generation after generation. The two pioneers of malarial discovery, the Frenchman Charles-Louis-Alphonse Laveran, and the Briton Ronald Ross, were both army physicians and both had been what we would call today "army brats." Laveran, born in Paris in 1845 but raised mostly in Algeria, was the son and grandson of military doctors. He graduated in medicine from the University of Strasbourg in 1867, joined the French Army as expected, only to suffer the indignities of capture by the Germans at the fall of Metz during the Franco-Prussian War. He was a prisoner of war for just a short time, then rejoined his regiment after being released. In 1878, at the age of thirty-three, he was posted to Bone, Algeria, where he was to create something less than a sensation by discovering the true causative organism of malaria.

The North African coast was malarious. There were species of anopheline mosquitoes that had adapted to breeding in the confined waters of oases; other species bred in the irrigation waters of coastal farms. The French introduced rice farming and the anophelines bred like flies. Later, much later, the colonial government, recognizing the error of its ecological ways, banned rice farming. In the clinic, hospital, and morgue Laveran was confronted, each day, with malaria. He became fascinated with the problems of its causation and pathology. Possibly, the few years he had spent teaching epidemic medicine (as his father had done before him) at the army medical school in Paris instilled in him the researcher's addiction for inquiry. In Bone, he was not your average "aspirin and calamine" army doctor attending to sick call. He managed to obtain a microscope. It was like the microscopes of his time, an optical toy.

On the morning of November 6, 1880, a feverish twenty-four-year-old soldier of the 8th Squadron of Artillery came to the office of Major Laveran complaining that he had been treated (probably inadequately) with quinine three weeks before and here he was, once again, with the ague. Laveran would have given the man another dose of quinine, but before seeing to the medication, on this day he took some blood from his patient and placed a drop on a microscope slide. Other doctors had examined the fresh blood of malaria patients (there were no staining techniques then), but Laveran saw under the microscope what no other researcher had seen. Laveran saw strange crescent-shaped bodies (the gametocytes of *Plasmodium falciparum*) and, even more astoundingly, some rounded bodies from whose surface filaments lashed and sinuously danced (the exflagellation of the male gametocyte). Laveran, at that moment, was the first ever to witness the parasite that caused malaria. In his (translated) words he was to write of that moment: "I was astonished to observe at the periphery of this body was a series of fine transparent filaments that moved very actively and beyond question were alive."[48] He also saw pale, ring-

48. Why Laveran saw exflagellation, the most active and observable characteristic of the malaria parasite, when no one else did is not known, but it may have been like so many other great discoveries, due more to serendipity than scientific logic. The male gametocyte will transform into gametes, exflagellation, at room temperature on a glass microscope slide as if it were inside an anopheline's stomach (fertilization may also take place but the process of development will go no further). However, exflagellation will occur only after about fifteen minutes. It may be that

like hyaline bodies within the red blood cells of patients with intermittent (malarial) fevers. It was a remarkable discernment; modern malariologists claim that Laveran's microscope was so optically defective it had only one half the magnification power necessary to reveal what he so accurately saw and described.

All these observations were described in a note he sent on November 23, 1880, to the French Academy of Medicine in Paris. It was met with disdain and disbelief. It was all so peculiar, so jumbled. Hyaline rings, crescents, dancing filaments. How could these be so many different morphological guises of one pathogenic microorganism? Nothing like it had ever been seen or described before by the established scientists of the day. Worse—Laveran had no credentials, he was a nobody from Bone. And it didn't help that Laveran was an atrocious artist. The drawings that accompanied his note were crude and unconvincing primitive art.

The Italians were particularly quick to dismiss Laveran's claims. They were, after all, *the* malaria arbiters. Malaria was theirs—the Roman fever; the bacillus of Klebs / Tomasi-Crudeli. The Italian scientists maintained that what this French upstart had described as the pathogen of malaria was nothing more than degenerating red and white blood cells. Laveran was just as certain that he had seen the true pathogen, and he persisted. Over the next year, working in Bone and Constantine, he collected a series of 192 malaria

other investigators looked at the fresh blood immediately and, seeing nothing, discarded the slide. Laveran must, for some reason, have delayed examining the slide for at least fifteen minutes. Why? What was he doing for those fifteen minutes? Treating his patient? Going to the bathroom? Neither Laveran nor his biographers enlighten us on the mystery of the missing quarter hour.

patients and in the blood of 148 of them he found the multiformed microscopic organism. This time, he wrote a 140-page report to the Academy. Neither his artistic ability nor his scientific credentials had improved. Laveran's thesis continued to be rejected. In 1882, Laveran went as an apostle to the heart of the matter—to the San Spirito Hospital in Rome whose beds were filled with those suffering from Mal'aria. Once again he demonstrates his rings, and round bodies, and crescents, and flagella-filaments, and once again the Italians reject what they are shown. Mal'aria is an Italian bacillus!

It was 1884, four frustrating years since he sent his malaria note to the Academy. In that year a new optical invention began to bring Laveran's parasites into sharp focus and began to convince the doubters of their reality. In 1884, Carl Zeiss, in Germany, designed the oil-immersion lens. With this lens, the magnifying power of the microscope was now trebled. Laveran's malaria microorganisms could be seen by all beholders. Nevertheless, there would be rejectionists for the next ten years. As late as 1887 the Russian zoologist Elie Metchnikoff (the father of immunology) took microscope slide preparations of malarious blood from Russia to the German Robert Koch (the father of medical microbiology) for Koch's seal of authenticity. The arrogant Koch made the anxiety-prone Metchnikoff wait outside in the hall for over an hour, then he cursorily examined the slides and declared that anyone who believed that this was the malaria parasite was a *dumkopf*.

By 1886 the Italian workers had finally accepted Laveran's organism as the cause of malaria and they began to dominate the important next step of "sorting-out" research. However, their irritation at having been beaten to the original discovery is reflected by the absence of any acknowledgment of Laveran in subsequent Italian publications.[49] Confirmation didn't bring immediate clarification. There was still considerable uncertainty as to the nature of the microscopic beast. Laveran had focused his attention on the filament-flagellum, which he thought was the end stage, the mature form of the parasite. Those lashing filaments somehow reminded him of primitive, microscopic algae and he gave the parasite a botanical name, *Oscillaria malariae*. The zoologists, however, correctly identified it as a protozoan, and Metchnikoff, who keeps weaving in and out of the malaria picture,[50] even placed it into the right general group of protozoa (the coccidia). Reason was on the horizon until the son of a Kharkov clockmaker threw malaria research into a state of confusion that would take almost ten years to clarify.

Up to this time (1886) everyone had been examining malarious humans for malaria parasites. Meanwhile on the

49. Laveran didn't do a great deal of research on human malaria after 1885, although in 1903 he founded a society for antimalarial work in Corsica. However, malaria had hooked him on the protozoa and he was to become a truly great protozoologist. The love of his later life were the flagellate protozoa (the trypanosome of African sleeping sickness is one such flagellate). He was awarded the Nobel in 1907 not only for his discovery of the malaria parasite but for his entire contributions on pathogenic protozoa. He used the entire 10,000 francs of the prize money for his laboratory at the Pasteur Institute—he had retired from the army in 1894 when they tried to make him an administrator. Laveran died in 1922. He was seventy-seven years old and was working on the flagellate protozoa in the sap of the cactus-like euphorbia plants. To the very end, he was the protozoologist's Protozoologist.

50. In 1960, when I was Professor of Medical Parasitology at the University of Singapore's medical school, the United Nation's World Health Organization sent me a first day cover envelope celebrating malaria. On the envelope were the pictures of those WHO considered to be the "Big 5" of malaria—Laveran, Ross, Grassi, Sinton, *and* Metchnikoff. The letter was sent to me addressed Singapore, China. Which tells you something about the World Health Organization.

steppes of Russia a zoologist-physician, Basil Danilewsky, M.D. (Kharkov) was, from 1884 to 1889, looking at the blood of birds and finding what seemed to be the same protozoan parasite as seen in human malaria. The Italian malariologists, taking a cue from Danilewsky, began to explore the possibility that malaria was a zoonotic disease; that the parasite was in birds and from them it was, somehow, transmitted to humans. "Proof" would be in the finding of infected birds in the same area endemic for human malaria. "Typical" parasites were found in the Spanish sparrow, the skylark, the chaffinch in Corsica, and owls and Italian sparrows in the Campagna. Malaria was a common infection of birds and people. And because some living parasites within red blood cells showed amoeboid movement, the Italians changed the name from *Oscillaria* to *Haemamoeba*—making it an entirely Italian parasite in the process.

It was all well and good to have a unitary concept of malaria, but in actual fact the Italians recognized that, clinically, human malaria assumed three distinct forms. There was *la terza benigna primaverde*—the malaria fever that occurred during the spring of the year. It produced forty-eight-hour fever peaks, and although the patients felt very ill they all made an uneventful recovery. Then there was *la febre perniciosa estivo autumnale*, a malarial fever occurring in late summer and autumn. It too had a forty-eight-hour fever periodicity but pursued an acute, frequently fatal course. Finally, there was a more uncommon, non-fatal malarial fever characterized by a seventy-two-hour interval between peaks. The question that the clinicians and pathologists raised was how could such a different trilogy of malarias all be caused by the same pathogen and how could that pathogen be common to birds and humans? In 1890–91 an untidy Russian Army pathologist was to provide the means by which the Italian malariologists could define and conform the multiplicity of malaria parasite species.

Dr. D. L. Romanowsky, the pathologist, was somewhat nonchalant in the management of his laboratory technique. He used the stains of his day, methylene blue and eosin, to make permanent stained preparations of tissues. Somehow, he forgot to replace the stopper on the bottle of methylene blue, and when he came to use it again it was mucky and abloom with mold. He used it anyway, counterstained with eosin, and to his amazement the cytoplasm of the tissues stained with a blue brilliance not yielded by fresh methylene blue, and the nuclei of the cells stained an intense red. The aging (oxidation) of the methylene blue had, serendipitously, turned it into a super stain.[51] The same polychromatic effect occurred when he stained a blood film from a malaria patient. The malaria parasites that could only be seen in vague outline in the usual unstained preparation now stood out vividly under the microscope—the blue cytoplasm and the red nucleus in the straw pink background of the host red blood cell. Now for the first time the parasites could be examined at high magnification, thanks to Mr. Zeiss and his wonderful optics, and compared for form, shape, and size, developmental stage by developmental stage.

Camillo Golgi, a professor of pathology at the University of Padua, clearly demonstrated that benign tertian and quartan fevers were caused by two different species of malaria parasites, each with their characteristic, diagnostic morphology. Golgi first noted the rapid streaming, amoeboid movement of the parasite of the benign tertian fever *(Plasmodium vivax)*, its active movement, as well as the rosette

51. The chemical magic that converts the methylene blue to azure as a result of oxidation and aging is still not understood with certainty.

formation of the quartan parasite *(Plasmodium malaria)*. The Romanowsky-stained films beautifully confirmed Golgi's descriptions. But the identity of the parasite that caused the malignant summer-autumn fever eluded him; and perhaps in frustration, Golgi left malaria and turned to the nervous system. It was not a bad career move because Golgi was awarded the Nobel in 1906 for his research on neurophysiology.

The man who was to become the most famous of the Italian malariologists, Giovanni Battista Grassi, took up the chase for the malignant tertian parasite's identity. Unfortunately, his opinion that it was a parasitic Trinity threw the issue into utter confusion. Grassi believed that the one disease, malignant tertian malaria, was caused by any one of three separate species of parasites, one of which was the same parasite as that infecting sparrows.

Finally, finally, in 1892 a collaboration of two Roman professors, the anatomist Amigo Bignami and the pathologist Ettiore Marchiafava, brought the Trinity to True Unity by proving that the crescents and rings (first seen by Laveran twelve years before) were all forms assumed by a single species of parasite. That parasite, after much taxonomic argument and revision, we now call *Plasmodium falciparum*, the cause of malignant tertian malaria—the death fever.

So, by 1895 the protozoan parasitic etiology of malaria was firmly established, although it was yet to be shown that the parasites enjoyed the privilege of sex. Nor, most crucially, was it known how the parasite got from A to B; how the infection was acquired and then transmitted from person to person.

Chapter 12

Man and Mosquito: The Briton's Tale

THE MIASMA THEORY of malaria continued to hang like a conceptual fog over the parasite's transmission route. At first, Laveran maintained a Hippocratic stance in his belief that the malaria parasite was contaminative—that it was spread through water and soil—and in vain he searched the waters and earth of Algeria for signs of the parasite. Meanwhile, a groundswell of hypothesis began to grow in favor of transmission by the mosquito. By 1884 even Laveran had been persuaded to the mosquito transmission hypothesis. All that was needed was experimental proof.

The idea of the mosquito as the vehicle of malaria sprang from epidemiological insights—guilt by association. Some observers of the malarial scene pointed out that in addition to the association of malaria, marshes, and miasma, there was, equally, an association of malaria, marshes, and mosquitoes. The natives often knew this and early travelers to Africa and Asia returned with amusing stories of how the natives told them that mosquito bites caused malaria. How ignorant those natives were is illustrated by the experience of the German colonists of German East Africa in the nineteenth century.

These colonists established farms in the healthy, non-malarious highlands. However, their entry into the country was through the intensely malarious port cities of the coast. It took them about two weeks to trek from the coast to their highland farm sites. This is almost the exact length of time that the malaria parasite, after the infecting anopheline has bitten, is silently developing in the liver (there are no signs or symptoms of malaria during this incubation period). Thus, soon after they arrived at their highland homes, the colonists came down with the first acute attack of malaria. This proved that malaria was caused by the change in climate they experienced as they went from the humid, hot coast to the cool-cold highlands, and not the mumbo-jumbo belief of the ignorant natives of the coast that malaria was caused by the bite of the *mbu* (Swahili for mosquito).

There were also Americans who espoused, without support of any experimental proofs, the idea that malaria was transmitted by the mosquito. In 1807, an Irish immigrant physician, Dr. John Crawford, published an article in the Baltimore *Observer* entitled "Mosquital Origin of Malaria Disease." Perhaps the most ardent advocate of the mosquito-as-the-carrier hypothesis was an American physician with the impressive name of Dr. Albert Freeman Africanus King. In 1882, King published an article in *Popular Science Monthly* in which he gave his opinion that the mosquito was the transmitter of whatever it was that caused malaria. An aside on the colorful, insightful Dr. King: he was at Ford's Theater the night Lincoln was shot and attended the dying President. King was not well versed in treating the traumas and wounds of emergencies since he was, at that time, Professor of Obstetrics at George Washington University.

One could imagine how a biting insect, such as a mosquito, might mechanically transport a pathogen, such as the malaria parasite, from one person to another. There would be a conceptual impediment, however, to imagining how intracellular parasites of warm-blooded animals would transfer to such an alien creature as an insect, where they would not only survive but undergo a complex developmental cycle that produced stages infective for the warm-blooded animal. It would boggle the mind that a parasite could perform such wild biological acrobatics.[52]

That some parasites of humans *must* undergo a developmental metamorphosis in an invertebrate was first demonstrated by a Russian, whose observations were augmented by a Scot in China, which were then brought to a conclusion, as a principle, by two Americans in Texas. In 1870 the Russian, A. P. Fedchenko, was the first to find the larvae of the guinea worm (*Dracunculus medinensis*) in the tiny fresh-water crustacean *Cyclops*. He didn't really know the significance of this, but research by later parasitologists revealed that infection occurred when a human swallowed the larvae-loaded *Cyclops*. In the human, the larvae turn into a 2-foot-long female and a 1-inch-long male (who fertilizes the female and then dies). The female wanders through the body for some time, then finally makes her way to the surface, creates a "boil," and discharges a myriad of larvae. The infected person, feeling considerable soreness about the boil, cools the lesion with water—usually water at the well in villages of Africa and Asia. The larvae then are ingested by the resident crustacea of the well waters. The native treatment is to extract the worm from the leg by

52. As they progress through their developmental cycle, passing from the human to the anopheline, the malaria parasites undergo a profound change not only in morphology but also in how they live, their chemical physiology, and their antigenicity. Neither the genetic mechanisms nor the molecular biology responsible for and governing these changes are well understood.

ever so slowly winding it onto a stick, maybe a turn or two a day. It is not uncommon in endemic areas to see people walking about with a stick fixed to their leg and a worm coiled on the stick. This is not malaria but it is, I think, an interesting digression.

By 1870 it had been proven that the elephant men (and women) of the tropics, those with the great and grotesque swellings of elephantiasis, were infected with threadlike roundworm (nematode) parasites collectively called *filaria.* It had also been shown that the adult worms live in the lymphatic tract and from there the female discharges several thousand microscopic larvae (the *microfilariae*) each day into the bloodstream. In 1870, however, it was not known how the filarial worms got from A to B; i.e., the way new infections began.

The Englishman, Patrick Manson, was twenty-two years old and already a physician when he became a Medical Officer of the Chinese Imperial Customs Service. The doctor-douanier spent thirteen years in Amoy, where he diagnosed and treated as best he could a panoply of exotic diseases of the Far East, including filariasis. Manson had caught the research bug. His curiosity led him to ponder the problem of filarial transmission, and when he returned to Amoy from home leave in England in 1876, he brought with him a brand-new microscope.

Manson, aware of Fedchenko's work on the guinea worm, and reasoning that the geographical distribution of filarial infections coincided with the range of mosquitoes, as well as with the unique circadian periodicity of the microfilariae (they appear in the peripheral blood only at night, a time when most mosquitoes feed), came to suspect the mosquito to be the carrier of the filarial worm. To prove his hypothesis, Manson fed mosquitoes on Hin Loh, his filaria-infected gardener. (This may have established an experimental precedence. Many years later, the former dean of the Bangkok School of Tropical Medicine employed a filaria-infected "sweeper" on whom mosquitoes were fed for experimental purposes and blood taken for student demonstrations. The man had an assured job so long as he remained infected and uncomplaining.)

The mosquitoes fed on Hin Loh were dissected on successive days and Manson saw under the microscope that the microfilariae not only survived but invaded the thorax and head, where they became larger and more differentiated. Manson had almost, but not quite, discovered the mosquito as the intermediate host of the filarial worm. He never did carry out the definitive experiments of infecting a human from the bite of a larva-carrying mosquito. Instead, he developed the hypothesis of the mosquito as a "nurse." Manson, who hadn't a clue about mosquito biology, believed that the female mosquito died almost immediately after laying her eggs in the water and went to her watery grave carrying the filarial larvae with her (actually, the female can live for weeks or longer after first laying eggs). People would become infected, according to Manson, when they drank the water containing the dead mosquito and the live larvae.

Manson returned to London, where he began practice as a tropical disease specialist. He became *the* specialist and because of his knowledge and charismatic character he soon gathered about him a following of young physicians who would be influenced by his ideas and become, when they returned to the tropics, the executors of his experimental proposals. One of those ideas waiting for a suitable executor was how the malaria parasite was transmitted. Manson extrapolated the filaria to malaria and proposed that malaria too was "nursed" by a carrying mosquito. For both the filaria and malaria, Manson got the basic mosquito concept right but missed, despite a generous clue provided by two Americans, the essential "in and out" of the transmission process.

In 1893, Theobald Smith and F. L. Kilbourne published their paper on the transmission of Texas fever of cattle through the tick. The causative organism, a *Babesia* (a protozoan parasite family relation to the *Plasmodium* of malaria), lives inside red blood cells in a manner similar to the malaria parasites. Smith and Kilbourne discovered that it was ingested by the feeding tick, and then *injected back* into another animal after a suitable period of development in the arachnid (ticks are like spiders; they are not insects).[53] This important observation established the principal that blood-sucking arthropods can act as active participants in the life cycle of pathogens. But Manson clung to the mistaken notion of the passive mosquito-carrying malaria. It was his belief that the malaria parasite developed into resistant "spores" within the mosquito. The mosquito drowned, the spores contaminated the waters, and infection was consummated by drinking the water. This misleading direction was to cause great confusion and difficulty for the "leg man" from India whom Manson snared in his research net—Surgeon Captain Ronald Ross.

Ronald Ross was the man most unlikely to succeed. Ignorant of zoology, he hardly knew one end of a mosquito from the other, let alone one mosquito species from another. He was opinionated, arrogant, and was to become eccen-

53. There is, however, a phenomenon in the tick-*Babesia* cycle that is absent in the mosquito-*Plasmodium* cycle. Without causing any harm, the *Babesia* parasite invades the egg of the female tick. Thus, the next generation of ticks are born already carrying the *Babesia* and ready to infect a new bovine host. Such vertical transmission does not occur in *Plasmodium*-infected mosquitoes.

tric in his single-minded pursuit of the malaria parasite in the mosquito. And he probably wasn't a very competent physician. Yet it was this man who, in July 1898, after years of wracking research, published his monumental discovery of how the malaria parasite was *really* transmitted by the mosquito.

Ronald Ross had no great interest in either the martial or medical arts. He considered himself to be an aesthete, a poet-playwright. Into his old age he continued to write the strained verse that reminds one of a late Victorian parlor. His father, a general of the British Army in India, decided that if his son wasn't going to be a soldier, then *certainly* he wasn't going to be a poet either. If he wouldn't fight he could, at least, be a doctor, and in 1875 he literally led his seventeen-year-old son by the hand to Barts (St. Bartholomew's Hospital and Medical School), the venerable institution located in the shadows of St. Paul's Cathedral and the Smithfield meat market.

Ross, an indifferent student, just managed to qualify as a physician by becoming what is considered the lowest rung of the legal medical ladder, a Licentiate of the Society of Apothecaries.[54] In 1881, now twenty-one, young Dr. Ross joined the military branch of the Indian Medical Service. Eight years later we find Ross overwhelmed by a kind of colonial cafard. His life has been directionless; the few hours

54. In Great Britain there were several ways of "skinning the examination cat" to become a physician. The smartest got a conjoint degree of Bachelor of Medicine, Bachelor of Surgery (M.B., B.S.). The next smartest received no academic degree but obtained the joint affiliation of Member of the Royal College of Surgeon, Licentiate Royal College of Physicians (M.R.C.S., L.R.C.P.). Those like Ross whose academic achievements were barely passable sat for the easiest qualifying examination, that of the Society of Apothecaries.

of work each day treating patients with familiar illnesses, the bachelor sporting life, and his own solitary pursuits of poetry and mathematics have not been sufficient. "My ponies browsed unsaddled, my books unread" sums up his mood of that time.

In 1889, he went on home leave with thoughts of abandoning medicine and India in favor of a poet's life. However, two events while on leave were to refresh and galvanize his life. First of all, he found a wife, Rosa Bessie Bloxam (about whom we know very little except that she was a constant sustaining force to Ross). No one has dared to inquire into Ross's sex life, but marriage must have ended a long drought of colonial celibacy. The "other ranks" were permitted their dalliances with the dhobie wallah's daughter and the brothels of the bazaar. Not so the bachelor officers, who hàd to sublimate their libido in such strenuous activities as pig-sticking and polo. The tidy English wives of the married officers kept a close and careful watch on the bachelors. A bit of "slap and tickle" with a native girl could easily ruin one's career in the Colonial Service. Acceptable, flirtatious affairs were provided by the covey of refined young ladies that the steamships brought to India each winter. They were, ostensibly, visiting relatives, but not having made it maritally in Britain they were hoping to be more fortunate amongst the unmarried—and presumably less discriminating but more ardent—men of the Indian Service. Fortunately for Miss Bloxam, Ross had evidently shown no interest in the "Fishing Fleet," as the group of visiting ladies were known to the bachelor officers. However, what revitalized Ross during the furlough even more than his marriage was the advanced training he acquired in public health and microbiology. He took the Diploma in Public Health course and then went back to St. Bartholomew's for two months of training in bacteriology. He now considered himself to be an authority.

To be authoritative, authorities need a special topic of their own. Ross chose malaria and was unwavering in his ignorance. You would have been an unfortunate patient if you had to consult Surgeon-Captain Ross for your attack of the ague. In 1893, thirteen years after Laveran's discovery of the malaria parasite, and seven years after the Italians' acceptance of the parasite's authenticity, Ross was adamant in asserting that malaria was an intestinal infection, probably caused by bacteria, that should be treated with calomel. Laveran, et al., were wrong—blood parasites have nothing to do with malaria. Laveran's parasites, illustrated by the most primitive drawings, were artifacts. In 1893, Ross was putting his pen to these opinions and publishing papers in the *Indian Medical Gazette* expressing his intestinal view of malaria. In 1894, Ross again went on home leave to London. There he met Manson and became a born-again malariologist.

From the beautifully stained specimens viewed under the finest optics of the day, Manson showed Ross what the malaria parasite *really* looked like. Ross became an instant convert to the plasmodial etiology, although Manson explained that the real mystery of malaria was not the pathogen but its transmittal. There followed frequent visits to the Master, and on a day in October 1894 Manson and Ross can be seen strolling down Oxford Street toward Hyde Park Corner. Manson is explaining his "mosquito-as-a-nurse" hypothesis of malaria transmission and urges Ross to pursue the parasite into the insect. Ross still has doubts as to his own future and is writing a romantic novel entitled *The Spirit in the Storm*, but by the end of the home leave he has been captivated by Manson's charisma and has become a disciple of research.

On March 28, 1895, Ross embarked, without his family, on the P&O steamship *Ballarat*. There was almost always at least one odd type amongst the passengers sailing to the colonies, and Ross must have been it for the *Ballarat*. He had purchased a microscope before leaving England, and with his new instrument-toy he insisted on examining the blood of his fellow passengers for malaria; he dissected the flying fishes that made suicidal landings on the deck; not even the resident roaches of the *Ballarat* were safe from Ross's microscopy. When he arrived in India, he was posted to Secunderabad, and there in a small regimental hospital he embarked on the long trail to follow the *Plasmodium* through the mosquito.

Ross's avocation of research made him an oddity amongst his fellow officers. Some years later, Knowles, who knew all about the standards of proper Indian Medical Service behavior, was to write from Calcutta about Ross in Secunderabad: "In 1895 a military medical officer who was keen on research was regarded as a somewhat unpleasant phenomenon—polo and pig-sticking being far more important." I recall one such officer in Nigeria who was enamored of dragonflies and was looked upon as the resident eccentric. Because of his tall, stooped figure and shuffling gait as he pursued the dragonfly, the villagers nicknamed him the *angalu*—the "vulture." Ross must have been the "dragonfly man" of Secunderabad as he sent his servants and hospital attendants out to collect mosquitoes and mosquito "grubs" (larvae), which he reared in jars to winged adulthood.

The servants, naturally, collected the most abundant, easiest-to-capture mosquitoes, which Ross, in his entomological naivete, classified only as "gray mosquitoes" and "brindle mosquitoes." The grays were probably *Culex fatigans* and the brindles *Aedes aegypti.* Both of these mosquitoes are a nuisance, but transmit certain viral and parasitic infections to humans, but neither is capable of transmitting human malaria or even sustaining partial normal development of the insect (anopheline) phase of the parasite.

For a year, Ross fed the gray and brindle mosquitoes on malaria patients with crescents (the gametocytes of *Plasmodium falciparum,*) in their blood. Day after day he dissected the insects in search of the "spores" which Manson predicted would be found in the "nurse mosquito." If you would be a Ross, try examining a cut-open mosquito under the microscope: total confusion to the untutored eye; the microscopic field crowded with bits and pieces of cells and tissues as well as microorganisms of the intestinal tract. How very difficult it would be to identify the developing malaria parasite amongst this profusion. However, from the day-upon-day dissections Ross gradually learned to recognize the normal cellular and tissue constitution of the mosquito. Under the microscope he observed the dance of exflagellation in the mosquito's intestinal lumen; but then there was nothing, no further sign of the parasite's progress.

Nevertheless, Ross adhered to Manson's thesis of the "nurse" mosquito/spore-contaminated water, and he acquired some natives (the history book calls them "volunteers") and had them drink a draught of "mosquito water." It was Ross's bad luck that one of these volunteers, a twenty-year-old Indian named Lutchman, actually came down with a fever eleven days after drinking water containing dead mosquitoes that had previously fed on a malaria patient. Ross was now certain that he had "malariated" Lutchman, although the fever had spontaneously disappeared and no malaria parasites could ever be found in the blood. Other volunteers were given "malariated" water with completely negative results. Lutchman was a coincidence—, although Ross (and Manson) clung to the notion that he was "special," that he got a "special" kind of malariated water, and that it was only a matter of a little more research to get the right "mixture" again. For the remainder of the year Ross pursued this false trail until finally Manson wrote him to give it up, that Lutchman was an anomaly. Ross should move on to pursue the flagella (the male gamete) elsewhere. So, following the advice of the Master, Ross pursued, day upon torrid day, the flagella, arising from the crescents, in the gray and brindle mosquitoes until "The screws of the microscope were rusted with sweat from my forehead and hands, and its last remaining eyepiece was cracked." Until August 16, 1897, all observations ended with the death of the flagella.

On August 16 a servant-collector, Mahommed Bux, brought Ross ten mosquitoes of a type that Ross had not noticed before. They were brown, dapple-winged, and had finely tapered bodies. When they fed, they put their heads down and their rear ends angled up in a straight line (unlike the feeding gray and brindle mosquitoes, the *Culex* and *Aedes*, which give a "hunchback" appearance while feeding). They were anophelines. The brown, dapple-winged insects were fed on a co-religionist of Bux, Husein Khan, a patient with crescents in his blood who lay ill with malaria in a ward of the Secunderabad Hospital.

Twenty-five minutes after all the mosquitoes had their blood meal on Khan, two were killed and dissected. Nothing; and then there were eight. Twenty-four hours later, two were found dead in the cage and two more were dissected. Nothing; and then there were four. On the fourth day, August 20, 1897, one had died a natural death and

Ross dissected one of the three remaining mosquitoes. There it was! A minute, round cyst on the exterior of the stomach wall. "The Angel of Fate fortunately laid his hand upon my head" was Ross's dramatic description of that moment. The last mosquito was sacrificed the next day. Not only were there cysts on the stomach wall but they were larger than those of the day before and had malarial pigment granules within them. They were alive and growing. By George, he got it!—almost. Now the Angel of Bad Luck and Confusion that had dogged Ross the past two years of research was to elbow aside the Angel of Fate.

Ross wrote a brief description of his findings and sent it off to the *British Medical Journal.* His note, entitled "On some peculiar pigmented cells found in two mosquitoes fed on malarial blood," appeared in the December 18, 1897, issue. Another batch of the mosquitoes with spotted wings and another few days would probably have given Ross the true picture of the malaria cycle, but this was not to be. First, Ross thought he saw it all, that the cyst was the end point. Manson was right after all in that the cysts must contain the "spores" of the malaria. Lutchman must have really contracted an infection from the "malariated" water that contained spores.

Second, Ross never quite completely grasped the principle that the human malaria parasite could develop *only* in the dapple-winged mosquito—the anopheles. This confusion was reinforced by the capture of a "red herring" mosquito. Ross was looking for more "dappled wings" but couldn't find any; anopheles mosquitoes are often difficult to find if you don't know where and how to collect them. But he did find a "gray" resting on the hospital wall and he dissected it. By sheer bad luck he found a typical cyst on its stomach wall. It must have been a *Culex,* a mosquito that could carry bird malaria but not human malaria. The chances were at least one in a thousand that that mosquito had fed on an infected bird, rested on the hospital wall, and was captured by Ross. Ross now thought that the grays as well as the dappled wings could carry the cysts of the human malaria parasite. To his later embarrassment, he published an addendum note giving this opinion in the *British Medical Journal*.

Nevertheless, he was still searching for more brown dapple-winged mosquitoes (why Ross during all his years of research never became acquainted with the fundamentals of mosquito taxonomy is a mystery) and on September 23 he found four and fed them on a malaria patient. He was ready to follow the flagella further, but the next day he received a telegram from the Surgeon-General ordering him to depart Secunderabad immediately for Kherwara in Rajputana, a semi-desert, barely malarious station almost one thousand miles away.

Manson was pulling strings at the Colonial Office in London to get Ross a special posting to continue the malaria research. Meanwhile, Ross was not playing the bureaucratic game by sending telegrams and demands to officialdom with the thinly veiled implication that they were a lot of incompetent fools. The poem he wrote when transferred to Rajputana reflects his contempt for the authorities:

> God makes us kings
> Scornful answer rings
> First be my scavenger . . .

Indeed, it is difficult to understand why the government wasn't interested in supporting investigations on malaria since one third of all hospitals beds in British India were occupied by malaria cases. However, on January 29, 1891, the "strings" finally tugged at the right places and Ross received orders deputing him to Calcutta on six months special duty to pursue his studies on the transmission of malaria.

Ross arrived in Calcutta on February 17 only to be asked by the Director-General when would he be prepared to leave for Assam to help in a kala azar epidemic raging there. And there were other problems: there were few malaria cases in Calcutta, and to make matters worse there were no "volunteers" for purchase. There was plague in Calcutta and the suspicious natives would not even allow a finger prick by the white sahib doctors (the Sepoy Mutiny and the brutal British reprisals were still fresh in the Indian memory) for fear that they might be given the disease. Taking blood for malaria diagnosis and then feeding mosquitoes on the positive cases was simply not possible during that plague year in Calcutta. Faced with this problem, Ross resorted to what all biomedical researchers must use when human subjects are not available, an animal model of the disease. Ross chose an avian malaria of the sparrow, which he correctly assumed to be biologically related to the human malaria parasites.

Within a month, working in his modest laboratory at the Presidency Hospital, Ross had defined his model system—sparrows infected with the parasite (then called Proteosoma) and the gray mosquito which was found to develop typical pigmented cysts on the exterior wall of the intestine a few days after feeding on infected birds.[55] On March 17, 1898, he fed the gray mosquitoes on an infected sparrow.

55. The sparrow Proteosoma was probably the avian malaria parasite now known as *Plasmodium relictum* and the gray mosquito was probably *Culex fatigans*. Unlike the human malaria parasites which only develop in *Anopheles* mosquitoes, the bird malarias use the *Culex* mosquitoes as their transmitting hosts.

Three days later, portions of the mosquitoes were dissected and the expected cysts found on the stomach wall. When the seventh day dawned, Ross fully expected to make the ultimate observation. This was the longest time he had let the parasite develop in the mosquito and by extrapolation there should now be fully developed spores within the cyst. The mosquitoes were dissected; the intestinal tracts placed under the microscope . . . and there was nothing! The bubble had burst. All Ross saw was the shriveled remnants of the collapsed and empty cysts. Careful search failed to reveal anything that looked like a spore or any other form that could be the transformed parasite.

Ross was frantic. What should he do next? He had less than five months to solve the problem to which he had devoted three years of his life. The whole edifice of his theory had collapsed with the empty cyst. Ross did what a lot of researchers still do when all experiments go wrong and cherished hypotheses go unsupported—we pack it all in and take a vacation. He took to the hills, to the cool beauty of Darjeeling, where for six weeks he luxuriated in the company of his family.

Refreshed, Ross returned to work in June. By the end of June he had timed a dissection to the critical day before the cysts (there usually are numerous cysts on the mosquito's stomach wall) burst. In the cysts of these mosquitoes he observed not rounded spores but rather a striated mass within the delicate, transparent cyst wall. June 29, 1898: Ross dissects a mosquito that had fed on a sparrow seven days before. The cysts (the oocysts are again empty, but this time he goes further with the autopsy and opens the mosquito's "chest"—its thorax. To his amazement, there in the thoracic cavity are not the expected spores but a myriad of microscopic, threadlike, thrashing bodies. They were the final transformation of the parasite. But how were they transmitted by the mosquito to the warm-blooded host? He searches the tissues but nowhere else can the "threads" (the sporozoites) be found.

It was July 4. It will be the holiday of Ross's final triumph. The mosquitoes to be dissected on that day had been fed on the sparrow twelve days previously. Under the microscope, Ross saw the shriveled oocysts, but the "threads" had disappeared from the thoracic cavity. He now performed a new kind of insect surgery. With the dissecting needle he carefully pulled the head away from the body. Attached to the head there were two ribbon-like pieces of tissue, and when these are put under the microscope they are seen to be packed with the "threads." What are the ribbon-like tissues? Other mosquitoes are beheaded and the course of the tissues followed to their entrance into the mouthparts. They are salivary glands! It was a breathless moment for Ross: after three years of hard labor the whole malaria transmission picture was revealed in that instant of discovery. He and Manson had been wrong. There were no passively transmitted spores; the malaria parasite transformed into these infective-stage "threads" (sporozoites) within the cyst and then invaded the salivary glands. Malaria was transmitted by the *bite* of the mosquito; by the injection of the "threads" when the mosquito took its blood meal.[56]

Ross immediately sent an account of his findings and mosquito-gland preparations to Manson in London and Laveran at the Pasteur Institute in Paris. Both of these men agreed that a discovery of the utmost importance had been made. A brief paper was published, and at the end of July,

56. This was proven several weeks later when healthy sparrows became infected after having been bitten by mosquitoes with "threads" in their salivary glands.

Manson announced the discovery to the British Medical Association meeting in Edinburgh. The entire meeting stood and cheered. This was Britain's heady Victorian period of adventure, and finding the mosquito transmission of malaria was much like finding the source of the Nile.

For Ross, the work was complete; the fate of the flagella had been followed to its ultimate destiny. In August, he left Calcutta for Assam and kala azar—a disease he considered to be nothing more than a pernicious form of malaria. He never took that final step to show that the human malaria undergoes the same cycle in the mosquito and that it is restricted to the "dapple wing" / *Anopheles*. He may have thought it superfluous to do so. He may still have been confused about the special, restrictive relationship between the malaria parasites of humans and the *Anopheles* mosquitoes. For this failure, the Italians were to put the "mosquito in the ointment." They were to maintain that Ross had *not* done it all—birds don't count!

Chapter 13

Man and Mosquito: The Italian's Tale

NO ROMAN HOLIDAY celebrated Ross's discovery. Once again the Italians had been upstaged by a nobody. First, they had failed to elucidate the causative organism of malaria. That discovery went to the Frenchman, Laveran. Now they missed the second grand prize of malaria research, the elucidation of the transmission cycle in the mosquito. That went to another amateur, this time a Briton. The gall was even more bitter because they (the Italians) knew that they had had all the best resources: the malaria was there, with mosquitoes and patients just down the Corso; the scientific expertise was there, Italy at that time had the most experienced malaria researchers in the world. Furthermore, many considered the leader of the Italian malaria research establishment, Giovanni Batista Grassi, to be the most outstanding zoologist of his time.

The Italians, as I have already noted, were continuously engaged in malaria research. Following Laveran's initial fundamental study, they were instrumental in tidying up the clinical aspects and parasitology-taxonomy of malaria. However, the problem of the transmission of malaria just did not, somehow, grab their attention or excite their enthusiasm. They carried out a few experiments in which an assortment of unidentified mosquitoes were collected in the malarious regions around Rome and allowed to feed on human subjects. When no infections resulted from the experimental feedings, this line of investigation was dropped almost completely. Certainly, no Italian pursued research on the transmission of malaria through the mosquito with the tenacity and single-mindedness of Ross. That is, they didn't pursue the mosquito until after July 1898, until after Ross had published his descriptions of the oocyst in the anopheline/dappled wing and the entire transmission cycle of the avian malaria through the "gray" culicine mosquito. The Italians, although they were later to deny it, were stunned by the reports from India and, under Grassi's direction, were hard at work by the end of July.

Giovanni Batista Grassi was born in 1854 in the Lake Como region of northern Italy. It is a region of wild and brooding beauty tempered by the fanciful pink villas and ancient villages strung along the lakeshore. The snow-covered Alps stand sentinel in the background and the 1200-foot-deep lake is surrounded by a wall of wooded mountains. In the spring the forests become carpeted, as if overnight, with green and white hellebores and the purple blues of violets and anemones. In the early mornings of late summer-early fall the Como people go to the forest to seek the edible wild mushrooms, particularly that great fungal trophy, the *porcini*. The close communion with their bosky surroundings tends to make amateur naturalists, applied biologists, of the Lake people, and some of their brightest have become professionals of the life sciences. But the citizens of Como are also an individualistic, somewhat contentious and competitive lot. Until the late eighteenth century, Lake town fought Lake town, although occasionally interrupting their rivalries to fight the common Milanese and Florentine enemy.

From early childhood, Grassi was fascinated with biology. He was to become a renowned zoologist and even today when one speaks of Grassi to zoologists, they are likely to recall not his work on malaria but his monumental studies on the metamorphosis and migratory pattern of eels. Grassi, like Ross, had no great interest in medicine and, like Ross, it was his father who directed him to the respectability of a medical degree . . . no peculiar Ph.Ds. for the Grassi family. However, the University of Pavia Medical School didn't cure Grassi of his love of natural history, and almost immediately after he graduated he went to Messina to study marine biology and protozoology, and then to Heidelberg where he wrote his thesis on bees. He was a great scholar, and a great scientist, but he was not all sweetness and light. If anything, he was as egocentric and contentious as Ross. Even today his supporters (mostly Italian) in the inevitable, bitter struggle for priority that erupted between the two men will sigh and admit that Grassi wasn't easy to get along with.

In 1895, Grassi was Professor of Comparative Anatomy at the University of Rome. For the past ten years his major research had been on the characterization of malaria of humans and birds. However, in July 1898 he turned to the problem of the transmission of *Plasmodium falciparum* by the mosquito. He collected *Anopheles claviger* (now known as *Anopheles maculipennis*), an anopheline abundant in the marshes about Rome, and fed them on patients with malignant tertian malaria. By the end of the year he had traced the course of the parasite through the mosquito and proved that the malarias of human are transmitted only by species of *Anopheles*.

The cycle of *Plasmodium falciparum* in *Anopheles claviger* was observed to be the same, in all major respects, as that described by Ross for the avian malaria parasite in the "gray" mosquito. Yet, when Grassi published his paper in November 1898, he cited Ross's work only at the very end, as a scant, grudging afterthought. In the paper published a month later, giving greater detail of his findings, Ross's publications are not mentioned at all.

Grassi, despite the importance of his findings, had come in second in the race. He was the Bishop of Malaria who confirmed Ross. But this was Rome and Grassi wanted to be Pope. Priority was, and still is, the first order of the business of science. Priority can sometimes be subtly manipulated by ignoring or dropping citations of the precedent work of others. Grassi was to maintain that he was unaware of Ross's work (although visitors to his laboratory at the time stated that they saw Ross's publication on Grassi's desk); that he had worked independently of Ross, and that his findings were of far greater importance.

Ross, as one can imagine, was infuriated. His sentiments were reflected in an editorial of the *Indian Medical Gazette* which stated that "Piracy in science is unfortunately nothing new."

The two big egos might eventually have come to some sort of amicable settlement had it not been for the intrusion of a third, even bigger, ego—Robert Koch. Although essentially a bacteriologist with little experience in protozoology, Koch was the final arbiter of late nineteenth-century microbiology. A stern, brilliant, proud, and patriotic man, who held the fatherland's problems as his own, he agreed to go to German East Africa on behalf of the government to investigate malaria transmission by the mosquito. On his return from Africa in 1898, he stopped off in Italy to do a few more "final" experiments. The Italian government gave every courtesy to the famous man who was charitable enough to help them with their malaria studies. Grassi was slighted and he burned with indignation. Koch's studies both in Africa and Italy came to nothing, and when Grassi completed his successful investigation he gave Koch "the needle" by triumphantly sending him a copy of the published paper. "I gave Koch a Christmas present," Grassi was reputed to have said.

Now everybody was outraged: Ross at Grassi, Grassi at Ross and Koch, and Koch at Grassi. Such is the conduct of science by the greats. But Grassi had made an unforgiving enemy of the German, and it was to cost him a share of the Nobel Prize.

There was no question in the judgment of the Nobel committee for 1902 that Ross should receive the Prize. The uncertainty was whether or not the award should be shared with Grassi; whether or not Grassi's work was of equal or near equal importance to that of Ross. The committee was leaning toward the shared Prize when Koch threw the full weight of his considerable authority in insisting that Grassi did not deserve the honor. Ross, still unassuaged, stood alone in Stockholm.

The arguments over who was the better man; who was the better scientist; whose work was of greater importance; and who should have priority for discovering the transmission of the malaria parasite through the mosquito has continued, like the aftershocks of an earthquake. In 1958, as a young scientist working in Africa, I attended the International Congress of Tropical Medicine and Malaria in Lisbon. On the way to a session I found myself sitting in a taxi between two renowned men of tropical medicine. On one side was Professor Saul Adler, an Israeli out of Britain and a former British Army medical officer during World War I. On the other side was Count Aldo Castellani, the scion of an old and noble Italian family. Castellani had carried out some of the early work on African sleeping sickness on behalf of the British and had been knighted by the king for his service to the Crown. That knighthood had been rescinded when Castellani forsook Britain in favor of Mussolini's fascism. Yet the heated argument between the two men was not about betrayal and fascism but about Ross and Grassi. Adler, ashes from his ever-present cigarette falling on his vest, was shouting at Castellani that Grassi was a liar, a thief of Ross's work. Castellani, then a very old man, was refuting Adler as best he could by maintaining that the Italians had made the truly important observations. And then I remember him saying, "It was the Germans, those damned Germans, who hated us Italians and denied Grassi the honor he deserved."

Thirty-two years later, on an unseasonably warm day in March, I was in the company of Bernardino Fantini, Professor of Medical History at the University of Rome. On our return to Rome from a day touring the formerly malarious Campagna and Pontine marshes, we were passing the Leonardo da Vinci Airport at Fiumicino. Fiumicino was now buildings and concrete; there was no evidence of it being one of the most malarious places in the world a mere sixty years ago. It was here that Grassi settled in his old age, attending to the sick, malarious children of the district. It was at Fiumicino that he died in the arms of his beloved daughter, Isabella, his last request being that she should look after his "children." Over the din of the jet traffic Fantini began to reminisce about Grassi and Ross, going over yet again the disputes and campaign for priority. Both had contributed so much to science and humanity. Both were honorable men who fell victim to the hubris of science. Fantini had the last word, and his remark as we left Fiumicino behind echoed that of Castellani—"It wasn't Ross; it was that damned German, Koch, who denied Grassi a share of the Nobel Prize and the recognition he deserved."

Grassi died in 1925, Ross in 1932. The traveler will find memories and memorials to them in obscure places. In Calcutta, near the old Presidency Hospital, a memorial gate bears the inscription:

> In the small laboratory 70 yards to the southeast of this gate Surgeon-Major Ronald Ross IMS, in 1898 discovered the manner in which malaria was conveyed to mosquitoes.

In Lecco, a small city at the end of Lake Como, on a busy one-way street near the railroad station there is a bronze street sign that reads:

Via G. B. Grassi
Scientziato 1854–1925

Chapter 14

"Treat the Patient, Not the Mosquito" —or Vice Versa

THE ROSS-GRASSI discovery of malaria transmission by the mosquito completed the essential knowledge of the parasite's life cycle. It was now possible to contemplate, in a rational manner, how malaria might be controlled. Unfortunately, each leading scientist-malariologist of the day had his own separate rational plan. This was particularly the case for Ross and Koch, stubborn, opinionated men who so permanently polarized the approach to malaria control that integrated, multifaceted campaigns have never been carried out except for rather halfhearted pilot projects. Ross, naturally, focused his attention on the mosquito. According to him, malaria could only be controlled by attacking the mosquito where it bred. Koch, loyal to the rapidly developing German pharmaceutical industry, was just as firm in his belief that malaria could only be drugged into submission. He disdained "swatting flies" as a primitive measure against insect-borne disease. "Treat the patient, not the mosquito" was Koch's dictum. Bold words from a one-drug doctor. That drug was quinine.

In the early 1900s, quinine was the only antimalarial drug. It was an antimalarial that had been around a long, long time, although the circumstances of its introduction into Europe are clouded in myth and mystery. The best-known story, although undoubtedly inaccurate, has a plot reminiscent of a 1950s Hollywood pseudo-history costume movie.

As that tale would have it, in 1638, Francesca de Ribera, the second wife of the Viceroy of Peru, was sick with malaria in Lima. Her physician, de Vega, is desperate to save the lady, and having no remedy in his pharmacopeia sends messengers throughout the land to seek a curative medicine. In the Andean town of Canizares (now in Ecuador) the natives had the bark of a tree that they claimed would cure malarial fevers. A packet of the bark was given to the intrepid messenger, who running night and day returns to Lima and with his last exhausted gasp gives it to de Vega. The dying countess is treated with a preparation made from the bark and a miraculous cure is brought about. The countess in her Catholic gratitude offers the drug to God via the Jesuits, who then carry it to Rome to dispense to malarious cardinals and popes.

The story may well be mostly fiction, but we do know that somehow the Jesuit missionaries in South America learned of the antimalarial properties of the bark of the Cinchona tree and that they had introduced it into Europe by the 1630s—there is reference to this in a chronicle written by a Father Antonio de la Calancha, published in Barcelona in 1638—and into India in 1657. By the eighteenth century a lively trade in Cinchona bark had been established between South America and Europe, where it was commonly known as "Jesuit powder." The problem, however, was quality control. Some batches of the Jesuit powder cured malarial fevers, while other batches had little or no potency. This variation was due, in part, to adulteration; fakery and quackery were common practices in the seventeenth and eighteenth centuries.[57] But the main reason for the variation in therapeutic effect, then unrealized, was that there were several species of Cinchona tree, each species having a different concentration of the active antimalarial alkaloid—quinine—in its bark. An adventurous young Briton, Charles Ledger, was finally able to identify *the* tree and in doing so influenced the destiny of the tropical world.

Charles Ledger came from a French Huguenot family who had migrated to England in the eighteenth century to escape religious persecution. In 1836, the eighteen-year-old Ledger was hired by an English trading firm to go to Peru to collect and export alpaca wool. The young Ledger became fascinated with the natural history of his new home and began to forsake the alpaca to journey through the mountains and forests. He also became intrigued with quinine, and in his travels he collected the seeds and bark of Cinchona trees. In one area he found a winner, a species of Cinchona that now bears his name, *Cinchona ledgeriana.* The bark contained the highest concentration of quinine that had ever been found.

The bark with its antimalarial bite should have found immediate acceptance in Europe. Much of Europe was still highly malarious. It was also a time of Europe's colonial

57. As a shell collector, my favorite example of fakery of the period is the Precious Wentletrap. The collection of marine shells from tropical oceans was as passionately pursued then as antique collecting is today. There were Sotheby-like shell auctions and rare specimens brought huge sums. One of the rarest and dearest was the Precious Wentletrap, a gorgeous, porcelain white Pacific shell with "flying buttress" ridges. Wonderfully contrived fakes were made of rice powder. Now, the Precious Wentletrap is known to be quite common and can be bought for a few dollars. The seventeenth-century fakes cost a small fortune.

expansion into the tropics—an expansion that was being hindered by malaria. Ledger collected the seeds of the tree in the expectation that their sale to the British government would bring him honors and fortune. He was so wrong—the British government spurned his offer. The desperate Ledger couldn't even peddle his seeds on the streets of London. The Dutch, then attempting to colonize Indonesia, were more perceptive. They bought Ledger's seeds and from them established Cinchona plantations in Java. Within a relatively few years the Dutch possessed the only assured source of high-potency bark and had a virtual monopoly on quinine. It was these assured and steady stocks of the Dutch-Javanese quinine that allowed the colonial penetration of tropical Africa. Nor did the Dutch forget their indebtedness to Ledger; they awarded him a yearly lifelong pension of £500.

By the early 1920s, quinine manufacture had progressed to produce a chemically pure preparation of predictable activity. It was such a potent antimalarial that there were those who regarded it as the magic bullet for malaria. Koch, the German ever looking for a chemical final solution, used quinine on populations of what was then German New Guinea. Here in this "hotbed of malaria" he believed that mass administration of quinine would make "malaria disappear for ever."

Quinine is, indeed, an excellent antimalarial and continues to be used extensively to save the lives of those stricken with strains of acute falciparum malaria, which is resistant to the more modern drugs. However, it could never make "malaria disappear for ever." Quinine is a short-acting drug that would have to be taken at frequent intervals for preventive (prophylactic) effect. But it can't be taken at frequent intervals because at effective dosage it has too many toxic side effects—ringing of the ears and, sometimes, deafness being the most prominent.[58] If malaria was to be conquered by drugs, a better magic bullet would be needed.

The ideal antimalarial would have to cure quickly and effectively; it would have to have a long-acting (prophylactic) effect; it would have to be free of untoward side effects; it would have to work against all species and strains of malaria parasites; and it would have to be cheap. Chloroquine came very close to that ideal.

Unlike the botanical quinine, chloroquine is a synthetically manufactured product. It belongs to a class of compounds known as 4-amino quinolines, antimalarials first developed in 1934 by the German pharmaceutical company I. G. Farben. The first 4-amino quinoline, Resochin, cured experimental malarias of birds and paretics, but it was considered too toxic for general use in humans unless given under close medical supervision. A few years later, the I. G. Farben chemists modified the formula slightly to form another antimalarial, which they named Sontochin. Sontochin produced few untoward toxic side effects, although its slow action in killing the malaria parasites was a problem.

In the late 1930s, I. G. Farben was the controlling power of a cartel that included Winthrop Stearns in the United States and Specia in France. The German "parent" informed both its American and French affiliates about Sontochin's antimalarial activity and how to synthesize the compound. Winthrop Stearns put this information on the "back shelf"

58. There is some evidence that tonic water contains a sufficient amount of quinine to have a slight, but measurable, antimalarial effect when drunk daily. The gin does nothing for malaria.

where it was forgotten.[59] Specia carried out human trials in North Africa and confirmed it to be the best, although not ideal, antimalarial then available. The Germans never continued research to produce a still more effective 4-amino quinoline because, it is said, Hermann Goering held the patent on Sontochin and was the sole beneficiary of the profits from the sale of that drug.

During World War II the American and Allied armies were engaged in battles in the malarious areas of North Africa, Asia, and the Pacific, where they were losing as many troops to malaria as they were to the Fascist and Japanese enemies. With the fall of Java to the Japanese, quinine, except for some old stocks, became unavailable. The United States very early in the war realized that a new synthetic antimalarial was urgently required, and a large pharmacological research program was carried out under the direction of the Board for the Coordination of Malaria Studies. Compound after compound was synthesized and empirically screened for activity against bird malarias. Compounds that showed some effect were then tested on experimentally infected prisoner volunteers at the Atlanta penitentiary. Over 14,000 compounds were screened, but only one proved to be acceptable for the treatment of human malaria. This was Atabrine (quinacrine), an 8-amino quinoline compound. It was a marginally active antimalarial that turned the skin a bright yellow, caused gastrointestinal dis-

59. In 1940 Winthrop gave a sample of Sontochin to malaria researchers at the Rockefeller Institute, where it was tested against bird malaria. It was reported to have a high antimalarial effect against the experimental model, but Winthrop did not exploit the lead or release the findings to other researchers. In 1943, when the wartime American expert malaria panel was reviewing potential antimalarial drugs, Sontochin was mentioned. The chairman of the panel, not being a chemist, mistook Sontochin's formula, a 4-amino quinoline, as being a more toxic 8-amino quinoline, and dismissed it for further consideration.

turbances, and, most alarmingly, occasionally caused temporary insanity.[60] Still, atebrine therapy was better than dying of malaria and during World War II it was the drug of choice—there was no other choice. The "seed" for the ideal antimalarial was Sontochin, which lay forgotten in the archives of Winthrop Stearns.

Finally, in 1943, Sontochin came into American hands. After the fall of Tunis, some Vichy French physicians who had carried out clinical trials with Sontochin gave samples of the drug to American military malariologists. Its chemical structure was analyzed in the United States, the composition changed slightly to give an even more potent therapeutic and prophylactic action, and renamed chloroquine.[61]

Chloroquine spread like a therapeutic ripple throughout the tropical world. In stately colonial homes, the bottle of chloroquine would be a fixture, along with the condiments, on the family table. In military cantonments the troops would assemble for "chloroquine parade."[62] In hospitals and rural health centers the malarious of all skin colors were treated with chloroquine and countless lives saved. However, there was never any serious consideration of

60. Atabrine was not even a product of the enormous wartime American antimalarial effort. None of the compounds synthesized by the American workers had any practical antimalarial activity. Atebrine, which was developed by the Germans in 1930, was resurrected and manufactured in America for military use.

61. When the chemical structure of chloroquine was compared to that of the "toxic" Resochin, they were found to be the same compound! The German researchers in 1934 were mistaken; Resochin was not too toxic for human use, and for almost fifteen years this ideal antimalarial had been forgotten and unused.

62. Two other antimalarial drugs became available after the war, paludrine and pyrimethamine. However, strains of malaria parasites resistant to these drugs appeared and they soon lost favor as therapeutics or prophylactics.

applying mass chloroquine administration—"drenching" the entire tropical world—as a strategy for the global eradication of malaria. The drug was not *that* cheap. There was no infrastructure to ensure that there would be widespread distribution, and even if there were, the opinion was that compliance would be poor; the natives could not be relied upon to take the prescribed regimen. And, in private conversation, colonial and national authorities would voice their concern that effective, widespread malaria control would open the Pandora's box of population explosion. In the 1950s, "the pill" was not yet available, and many felt that there was no alternative to the cruel culling of 30 to 40 percent of the children by malaria to maintain population, cultural and economic stability in the tropics. Then, in the 1960s, even the thin, saving remedy of chloroquine was lost.

During the early sixties there were ominous reports from South America that chloroquine was not curing falciparum malaria. It was soon confirmed that chloroquine-resistant strains of *Plasmodium falciparum* had arisen under the pressure of overusage and, probably, underdosage. Slowly but inexorably, drug resistance spread throughout the entire malaria-endemic regions of the world—to South and Southeast Asia, Africa, and Melanesia. Geographical isolation was no barrier; chloroquine resistance invaded the most remote areas. After thirty-five years of a malaria-free life in the tropics, I came down with an attack of chloroquine-resistant falciparum malaria after a two-night stay in a New Guinea village that was three days away from the nearest motorable road.

There was no drug to treat chloroquine-resistant malaria except that ancient antimalarial, quinine. But there was hardly any quinine available. During the twenty years that chloroquine held therapeutic supremacy it had almost totally replaced quinine. The Cinchona plantations of Java, long unprofitable, were replanted with other crops. During the Vietnam War, warehouses were combed and very old bottles of quinine, their labels browned and crumbling with age, were treasured finds. In more recent years Cinchona plantations have been reestablished. Also, quinine has now been completely synthesized. A synthetic analogue of quinine, mefloquine, has been developed, but until recently it has had a parsimonious distribution because malariologists feared that overuse of this drug would lead to strains of *Plasmodium falciparum* that would be resistant not only to mefloquine but also to quinine. If resistance arose to the last and only effective antimalarial, it would be a disaster—a slaughter of the innocents.

No major efforts have been made to find an antimalarial to replace chloroquine. The motivations of colonialism and profit are gone. Pharmaceutical companies that once pioneered the development of drugs to treat malaria and other tropical diseases have, for the most part, dropped this line of research. The now astronomical costs for chemotherapeutic research, experimental and clinical trials, and steering any new drug through the legal-bureaucratic maze to obtain patents and FDA approval, have made the development of drugs to treat the diseases of poor people uneconomical. The stockholders would never approve of such altruistic fiscal irresponsibility.

The best antimalarial hope on today's horizon is a "new," two thousand-year-old drug called Qinghaosu. The Chinese, whose southern regions have always been malarious, were not prepared to wait for the new antimalarial drugs that never emerged from Western technology. Moreover, in the 1960s the Chinese Communists were turning inward; Western thought—including the Western system of medicine—was rejected as wrong and corrupting. In 1967, the Chinese, under Mao, began a systematic search of their ancient medical manuscripts for remedies to expand their traditional herbal pharmacology. In a book written by Ge Hang in A.D. 340, *The Handbook for Emergency Treatments*, mention was made of the remarkable febrifugal properties of the sweet wormwood (*Artemisia annua*) in the treatment of periodic fevers. Heeding the advice of Ge Hang, the modern Chinese pharmacologists collected Artemisia plants, prepared a decoction from the leaves, and tested it for an antimalarial effect on mice infected with a lethal malaria (*Plasmodium berghei*). Its activity in the mice was as good as that of chloroquine. It also cured chloroquine-resistant strains of the mouse malaria.

The plant extract was further purified and in this form (artemisinin) tested in human patients with malaria. It was found to cure falciparum malaria of humans more rapidly and with less toxicity than chloroquine or that other ancient herbal—quinine. It has been effective in treating the most deadly form of falciparum malaria—cerebral malaria. It has been effective against strains of *Plasmodium falciparum* that are solidly resistant to chloroquine. Now chemically analyzed, it could be synthesized for large-scale production. However, if you are planning a trip to a country where there is chloroquine-resistant malaria, don't go to your druggist and ask for Qinghaosu. He hasn't got any. For reasons that are not clear, it is not being produced by any Western pharmaceutical company. The malarious world desperately needs a "new chloroquine" such as Qinghaosu. Hopefully it will not require the cataclysm of epidemic or war to bring the new antimalarial to your druggist's shelf and the shelves of the primary health centers wherever chloroquine-resistant *Plasmodium falciparum* is a killing predator.

However, as we noted earlier, it has never been possible to achieve the great goal to expunge malaria from the world by means of chemotherapy. Was Ross, the advocate of mosquito control, correct and Koch, the chemotherapist, wrong? After he left India, and left research, Ross traveled to many tropical countries preaching the necessity of mosquito control. He was now an important personage, a knight and a Nobel, but he still seemed ignorant of mosquito biology. He thought it a simple matter to find the mosquito "grub" in its watery lair and to suffocate it by engineering works that would remove the water, or by oiling the water's surface. An accomplished mathematician, Ross had the figures to "prove" how malaria eradication could be achieved by mosquito reduction. He simply would not heed the counsel of seasoned malariologists, who recognized the enormous diversity of anopheline behavior from species to species and, therefore, the enormous problems entailed in their control.

During the first third of this century, the wealthy Western countries, motivated by prospects of economic self-benefit, carried out massive antimalarial engineering works. America needed the Panama Canal, so General W. C. Gorgas drained the swamps and seepages to rid the Canal Zone of malaria and yellow fever-carrying mosquitoes. Great Britain needed rubber for its growing automobile industry and Dr. Malcolm Watson in Malaya devised subsoil drainage to destroy anopheline breeding sites, thereby keeping the plantation coolies healthy and tapping. Perhaps the most monumental of the antimalarial engineering projects—bonification—was carried out in Mussolini's Italy.

We now remember Mussolini as a posturing buffoon whose life ended strung up by his heels in a village of Grassi's Como district. It was not always so. In his time many Italians, including prominent malariologists, condoned his politics and policies. He not only made the trains run on time but he removed the ancient burden of malaria from Rome to the sea. For this, Mussolini is still remembered by the Italians of the Pontina with gratitude and honor.

The Campagna begins below the hills of Rome and continues west to the sea as the Pontina. The Appian Way, Imperial Rome's lifeline to the sea, ran through the Pontina. The Pontina is a wetland but varied in its wetness. There are marshes; forests within which stand pools of water and ponds; small lakes and soggy grasslands. There were, also, anopheline mosquitoes, and the Pontina was one of the most malarious places in the world. Each day shepherds would come from the hills to tend their flocks of sheep and return to the safety of the heights in the evening. Water buffalo thrived in the marshes. Italy's finest mozzarella came from the Pontina buffalo and the pecorino cheese from its sheep. In the 1920s, eucalyptus trees were planted in the Pontina in the mistaken belief that they emitted a volatile substance inimical to mosquitoes. With or without the eucalyptus trees, the Pontina remained malarious and uninhabitable.

During the 1930s, Mussolini, heeding the advice of Italian malariologists, undertook a bold and costly project to make the Pontina safe for human life. An extensive complex of drainage canals, the largest being the Mussolini Grand Canal, was constructed. The Pontina began to dry and the anophelines lost their breeding sites. Malaria transmission was reduced to a point where human settlement became possible. Farms were built, each with a red barn, and given to World War I veterans. Pontina towns abandoned by the Romans almost two thousand years earlier came to life again. Prosperous new towns, such as Sabaudia, rose from the formerly malarious marshes. In the Roman Campagna an apartment city and the international airport at Fiumicino came into being. Tablets and memorials praising Mussolini can still be seen in the towns of the Pontina. In Sabaudia there is a large memorial pillar erected in 1938 whose inscription gives effusive homage to Mussolini for rescuing the Pontina from "a thousand years of death and sterility."

The Pontina was drained and malaria controlled, but not completely eradicated. Transmission didn't come to a complete end until about 1948, when environmental pollution further reduced the anopheline population. The vector, *Anopheles maculipennis,* demands pristine waters for its breeding sites. Waste from the revitalized Italian industries and human waste from the growing population in the Pontina killed off most of the remaining anophelines. Ironically, the environmentalists are now pleased to find anophelines in the Pontina. The clean-living *Anopheles maculipennis* is the new sentinel; its presence is testimony that the antipollution measures are successful. Nor has everyone approved of making the Pontina malaria-free. As malaria went, so did the mozzarella. The water buffalo almost disappeared from the dried and populated Pontina.

Although the impressive, large-scale works of Mussolini in the Pontina, of Gorgas in Panama, and of the Rockefeller Foundation in Sardinia were undoubtedly beneficial, they did not help the great mass of people at risk from malaria—the agricultural and jungle populations of Africa, Asia, tropical America, and Melanesia. Big environmental engineering projects cost big money. For the impoverished Third World, an inexpensive, relatively simple, relatively long-lasting antimalarial measure was needed. That need was met by the discovery of DDT; and for the first time in human history there was a weapon not only to control malaria but, potentially, to eradicate it for all time from all places.

DDT is uniquely long-lasting—its residual effect is so powerful that a single spray application on a house wall will kill mosquitoes up to six months later. Because DDT had an extended activity, the neo-Rossian malariologists could look at the epidemiological mathematics of malaria again and formulate a new plan for eradication—"the permanent concern of the State," as the Master said in 1911. The logic of that malaria eradication program was based upon a number of postulates. These were: (1) The important malarias of humans have a finite course of infection. Some of the untreated may die during the first attacks, but the survivors will go to complete self-cure, if not reinfected, in about two years for one species of malaria parasite (*Plasmodium falciparum*) and approximately five years for the other common malaria (*Plasmodium vivax*). (2) The main vectors of malaria are (or were) anopheline mosquitoes that fly into dwellings in the evening to take their blood bites and, having done so, take a postprandial digestive rest on the house walls. (3) Ergo, if every wall of every house in the malaria-endemic area were sprayed with DDT every three to six months, there would be no further transmission of infection because (a) the mosquito population would be reduced to so small a size that transmission would be (by mathematical projection, at least) a near impossibility, and (b) the DDT-affected mosquitoes would live too short a time for their parasites to complete their life cycle to the infectious stage.[63] (4) If these DDT spray cycles were carried out assiduously for five years, over which time there would be little or no transmission (Postulate #3), then all the malarious people would become "burnt-out" cases (Postulate #1) and there would be no more malaria. Everyone could then relax and all that would be left would be some minor mopping-up operations to detect and treat the few "escapees." A country could become eternally malaria-free if it had the resolve and willingness to divert the greater portion of its health resources to an eradication program for a period of five years.[64]

Pilot projects to control malaria by DDT spraying were carried out in the Tennessee River Valley, Cyprus, Greece, Venezuela, Guyana, and Puerto Rico. The results from these pilot campaigns were so good that as early as 1950 senior malariologists were advising to "go for it!"—global eradication of malaria. In 1955, the World Health Organization Assembly, assured of success by the mathematical projections of the British epidemiologist George MacDonald, and by the promise of large amounts of money from the United States, endorsed the policy of global eradication.

Malarious nations throughout the world joined the crusade. And they contributed their human and material resources. Poor countries whose entire annual per capita health budget was $1 would allocate 35¢ of that dollar to

63. It takes about fourteen to twenty-one days from the time the anopheline mosquito takes its blood meal from one malarious human until, after a series of complex transformations, the parasite becomes the infectious (sporozoite) form in the now "loaded" mosquito's salivary glands. Thus, the DDT-affected mosquito, even if it took a blood meal of parasites, would live too short a time for it to become a transmissive menace.

64. The anopheline population would increase again after the spray campaign, but now they would only be nuisance mosquitoes. There would (theoretically) no longer be any human infections to act as a source of infection to the mosquitoes.

the malaria program. From 1956 to 1969, the United States, through the U.S. Agency for International Development (AID), gave $790 million to the Global Eradication of Malaria Program. By 1969 it had become apparent that all that money and effort had gone down the drain, that the global eradication of malaria had been an impossible dream.

There were a number of reasons for the failure, not least that the anopheline vector mosquitoes were becoming resistant to the action of DDT both physiologically—they developed the enzymes to detoxify the insecticide—and behaviorally—instead of feeding and wall-resting, they changed in character to feed and then quickly bugger off to the great outdoors. There is persuasive evidence that the antimalarial operations did not produce mosquito resistance to DDT. That crime, and in a very real sense it was a crime, can be laid to the intemperate and inappropriate use of DDT by farmers, especially cotton growers. They used the insecticide at levels that would accelerate, if not actually induce, the selection of a resistant population of mosquitoes. Human resistance was also developing. Led by the Silent Springers, there developed a revulsion to all things chemically insecticidal. It was not recognized that DDT used for medical purposes never killed an osprey. It was the intemperate use by agriculturists that caused the insecticides to spill over into the environment.

Then, too, there were anopheline species whose natural history had never been reckoned with by the eradication planners. These were species whose genetically programmed behavior directed them to remain outdoors. They never came indoors to feed and never, never came in contact with the DDT-sprayed walls. In addition, there were species whose biological efficiency in acquiring and transmitting the parasite and whose reproductive capacity were so great that no amount of DDT alone could ever cause the interruption of transmission. For example, computer analysis of the epidemiological-epidemiological data from Africa revealed that to interrupt transmission by the incredibly efficient vector *Anopheles gambiae*, 50 percent of the entire anopheline population would have to be killed off *each and every day* of the five-year attack phase of the eradication program.

The World Health Organization never fully acknowledged its responsibility for the global program's failure. It never fully acknowledged that it had miscalculated in planning operations against the "awkward" mosquito vectors. Nor did it admit that it had oversold the program to the Third World and that it had set unrealistic, if not impossible, goals. It did not acknowledge that its staff were not always aware of the true state of affairs. A malaria expert from WHO headquarters in Geneva would make an advisory inspection trip to the Third World country. He would stay at a hotel in the capital city. He would be taken for a visit to the field for a day or so, return to the hotel, speak to the officials who gave him "official" figures. He would then return to Geneva, where he could refresh himself from the burden of travel with a meal at Aux Fin Bec. Complaints from the national program's "field hands" that WHO's program design could not be adhered to and that control and not eradication was the only realistic possibility were viewed by the WHO officials as apostasy. Unfortunately, the apostates proved to be correct.

By 1967 it became apparent, even to WHO, that the global eradication of malaria was not possible, and the tune was changed to "control." This was the last song that was sung. The sixty-three countries that had enrolled in the eradication program and had put so much of their scarce resources into it decided to call it quits. It was one thing to make a short-term capital investment and then never have to pay another cent. However, "control" was another matter. It meant that these poor countries would have to make an endless payment to malaria. The spray teams would have to be employed indefinitely—men, vehicles, insecticide, support services. In addition, WHO now recommended adjunctive distribution of antimalarials. Funds from donor countries, notably the United States, were being withdrawn—"Success has a hundred fathers. Failure is an orphan."

In 1972, the Global Eradication of Malaria Program was formally declared dead, and expert-bureaucrats began the postmortem assignations of blame. The apparatchiks of WHO covered their derrières by laying the main blame on the national workers, who were accused of not having performed as directed. This was a diplomatic way of saying that the natives were too lazy and too ignorant to have done the job properly. USAID technocrats joined WHO in this assignment of blame, and added some face-saving economic sophistry.

Economic self-interest has long been a powerful motivation for industrialized countries to give money to improve the health of the Third World. Healthy people produce more, thus make more money, and thus can afford to buy Chevrolets and Frigidaires (although now in the Third World anyone with money—even American-donated money—would more likely buy Toyotas and Sanyos). It could equally be argued that given a fair price for their produce and labor, people will have enough money to lead a healthy life style and afford health services.

AID had used similar arguments in obtaining the money and confidence from Congress for underwriting the Global

Eradication of Malaria Program. An AID apologist when that program failed was not a malariologist or anyone even remotely associated with malaria but an economist, Edwin J. Cohn, of the AID Office of Policy Development and Analysis. He contended that not only didn't the failure of the campaign matter but it may even have been a blessing in disguise. The Third World didn't require a healthy labor force because there was a surplus of workers; better, some people should be sick with malaria and spread the job opportunities around. He also said in effect, on behalf of AID, "better dead than alive and riotously reproducing." This was malaria and the Big Bang all over again: freedom from malaria lowered the crude death rate with sustained high fertility, which caused the rapid population growth, which led to the decline of economic development. The Third World would be well advised to return to the condition it enjoyed before embarking on the campaign to eradicate malaria.

However, during the ten years that the program was in effect, the nature of malaria had changed. There was no going back to pre-1955 conditions. The parasites had become resistant to the effective, inexpensive antimalaria drugs and there were no new therapies to replace them. When the Vietnam War ended, the U.S. Army's chemotherapeutic research effort, one of few programs trying to develop new antimalarials, began to wind down.

The biology of the anopheline vectors had also changed during the "eradication years." Many species became resistant to insecticides. Important vectors had changed their behavior. The degradation of tropical ecosystems affected anopheline-human relationships. In Africa, for example, malaria had been a rural disease carried by "country" *Anopheles gambiae*. The big cities of tropical Africa had been relatively free of malaria because their "concrete-slum" ecosystem provided few suitable larval-breeding habitats. The wholesale destruction of the African forests has led, through selective pressures, to the replacement of the forest-dwelling type of *Anopheles gambiae* with a variant that is adapted to thrive in the cities. The teeming cities of tropical Africa have become increasingly malarious.

Some predict that the greenhouse effect brought on by environmental abuse may make our own temperate zone malarious once again. J. D. Gillett of the London School of Hygiene and Tropical Medicine warns that the increase of pollution-generated carbon dioxide in the atmosphere will cause a global warming in which the average minimum temperature of the temperate zone will rise to 27.5°C (81.5°F). The present temperate zone would become subtropical—and the efficient tropical malaria-transmitting mosquitoes would move in. Immigration of people from the Third World would ensure a continuous reservoir of infection for the advancing anophelines.

There has been no new consensus as to how malaria should now be brought under control. The World Health Organization, together with the Food and Agricultural Organization and the United Nations Environmental Protection Program, have recently recommended environmental measures, chiefly water resource management, as the best long-term method to control malaria. They pointed to the Al Hassa Oasis project in Saudi Arabia, where concrete drainage canals were constructed at a cost of $84 million. The breeding of the malaria vector, *Anopheles stephensi,* was halted and agricultural production was also vastly improved. But this example would hardly apply to most nations of the Third World without Saudi Arabia's oil-rich ability to make an $84 million capital investment in a water-management project. More importantly, many of the tropical malaria-transmitting anopheline species of the tropics breed in small collections of water such as puddles and pits which cannot be "dried out" by canals or other engineering works.

There are experts who recommend a return to an antimalarial technology of an earlier time—the mosquito net. Mosquito nets are portable, cheap (they can be made for $2–$5), and when used properly they are a remarkably efficient way of reducing malaria transmission. The trouble is the natives too often don't use them properly and they don't like them because they are too hot to sleep under—a mesh fine enough to keep out the mosquitoes will also keep out the air, and a mosquito net that is in disrepair or not properly tucked in will trap mosquitoes rather than bar them. Too often there aren't *enough* mosquitoes to persuade people to buy and use a mosquito net. Villagers will take to sleeping under a net to protect them from the swarms of non-anopheline nuisance mosquitoes to get a good night's sleep. They are less likely to buy a mosquito net as an antimalarial measure, and in most places anopheline attack is light enough to allow an undisturbed sleep.

Perhaps the most promising method of dealing with malaria is a simple technology that combines the mosquito net with the insecticide. There is a quite new, rapid "knockdown" insecticide called permethrin which is non-toxic and has an active life of several months. Mosquito nets that have been dipped in a permethrin solution have been shown to repel and kill mosquitoes for a month or more. The mosquito nets can also be made with a larger mesh and allow for better ventilation. Pilot studies in which villagers have been given permethrin-dipped nets, or as a community endeavor dipped their own nets in a common 44-gallon drum containing a suspension of permethrin, have shown a significant reduction in malaria.

Finally, an influential body of opinion maintains that simple and cheap methods, such as the bed net, are not an appropriate way for a world about to enter the twenty-first century to deal with a problem such as malaria. These experts maintain that biotechnology will save the day through a new "magic bullet." It was the appeal of the biotechnological quick fix that brought AID back into the malaria business. It was showtime again. In 1965, AID began its $100 million quest for the ultimate antimalarial—the Vaccine.

Chapter 15

The First Trials and Tribulations of the Malaria Vaccine

THERE ARE SCIENTISTS in America, Australia, Great Britain, Colombia, and undoubtedly elsewhere who fall asleep murmuring "Thank you, thank you"—in Swedish. They dream of the Prize that will be theirs when they discover the malaria vaccine. There are also other malaria vaccine researchers whose present reality is a cell biology of another sort. One is in jail. One is about to go on trial, indicted for theft, criminal solicitation, and conspiracy. Two others have been indicted for theft. Others are under investigation and may or may not escape cell time. Corruption may bring an elusive but worthy goal, pursued for almost one hundred years—the vaccination against malaria—to a tawdry and inconclusive end.

Laveran and the early malariologists did not conduct their research in an isolated desert of knowledge. The 1870s to the early 1900s were the gorgeous opening decades of microbiology and its companion science of immunology. In 1887, Louis Pasteur, by experimental mismanagement, found that inoculation of an aged, dying culture of *Pasteurella antiseptica* into chickens would subsequently prevent the birds from acquiring fowl cholera, the killing disease normally caused by this bacillus. The birds were resistant to challenge with a virulent organism; they had been made immune. This established a basic principle of immunization with vaccines: a microbial pathogen modified or manipulated in a way that renders it harmless induces a protective immunity in otherwise susceptible animals and humans.[65]

Pasteur's protection of the chicken was the opening shot of immunology's most practical endeavor, the production of vaccines to protect against infectious diseases. In the hundred years since Pasteur's initial observation, vaccine after vaccine has become available to protect us from a great variety of infectious disease—anthrax, rabies, tetanus, diphtheria, pertussis, yellow fever, polio, and many more. Solid protection from a shot, or series of shots, in the arm has been gifted us from the microbiological, pharmacological, and now biotechnological laboratories to build an ever-widening immune fortress in and around our vulnerable bodies. It is no wonder that from the late nineteenth century onward, clinical medicine and its neglected stepchild, public health, looked to vaccination as the ultimate solution to each and every infectious disease. It is such a seductive stratagem; everybody in the whole wide world could be immunized against the panoply of viral, bacterial, fungal, and parasitic pathogens from an early age. Drugs—che-

65. Pasteur was not the inventor of vaccination. That discovery belongs to Edward Jenner, who in 1778 inoculated workhouse children with cowpox and then challenged them with pus from the pustules of smallpox cases—see Desowitz, *The Thorn in the Starfish: The Immune System and How It Works* (New York: W. W. Norton & Co., 1988). Unlike Pasteur, Jenner had no idea of the microbial (viral) nature of smallpox; he came to vaccination by thoughtful clinical and epidemiological insight. That first vaccination has proved to be the best vaccination. The smallpox vaccine renders immunity for life, and since there are no smallpox variant viruses, it is protective throughout the world. Because of these unique characteristics it was possible, in the 1970s, to achieve global eradication of the disease.

motherapy—while useful to treat the unimmunized "escapees," would be required only as a second-line armament held in reserve. And it almost goes without saying that this vision has to a large extent been fulfilled. At least the *potential* is there to immunize populations, given the will, strategy, and money. There is even the possibility of actually eradicating certain infections for all time from all places by immunization. Smallpox was first; measles may be next. But there are notable exceptions to protection by vaccine—the common cold and AIDS. And malaria.

Thus, the post-Laveran malariologists were struck with the notion that "their" disease could be prevented by immunization as readily as the bacteriologists who were so successfully vaccinating against "their" diseases. Certainly, during the almost half-century span between 1900 and the end of World War II there was persuasive reason and motivation to try to make a malaria vaccine. The cheap, potent, relatively long-lasting antimalarial drugs were syntheses of the future. Only the short-acting, dizzy-making, ear-ringing quinine and its analogue-derivatives were available for treatment and prophylactic prevention. And even quinine was denied when the Cinchona plantations were consumed by the Japanese invaders of Java in the opening years of World War II. Nor were the wonderfully efficient insecticides of the postwar years, DDT and its chemical successors, then available to control transmission by the mosquito vector. There were only bed nets for the affluent individuals, and laborious, expensive environmental works, such as draining a mosquito-breeding swampland, for the affluent nations.

The motivation to pursue research on a malaria vaccine was not so much altruistic as it was imperial. Colonialism by the Great and Not-So-Great Western Powers was firmly entrenched by the end of the nineteenth century. The concern for the health of the native populations, beyond maintaining a reasonably healthy workforce for commercial enterprises, was subordinate to the concern for the health of the colonial administrator, his military support, and the settlers and traders from the ruling nation. The perpetuation of the colonial system depended upon keeping the administrative, military, and economic rulers alive and healthy. Malaria was always a constant threat to the viability of the system, and in some places it almost made the "mother country" lose her grip. The early days of exploitation by Britain, France, Spain, and Germany in West Africa are one such example—"The Bight of Benin. The Bight of Benin. Few go out but many come in," as the nineteenth-century lament had it.

World War II brought an additional need for a better means to protect troops from malaria and gave renewed motivation to pursue research on the malaria vaccine. The Allies were losing men to malaria in the South Pacific, in Asia, in Africa, and later, in Italy.[66]

For the invading armies of the nineteenth and twentieth centuries and the colonial administrators who followed in their footsteps, there was little that had changed from the threat of malaria since the time that Sennacherib and his Assyrians came storming down the Jordan Valley, or Napoleon lost his men in 1798 as they were camped near Acre and lost them again to malaria (70,000 cases, 10,000 dead) in 1809 during the Walcheren Island campaign. The

66. The Axis Powers were having their malaria troubles too, although as far as we know their scientists did not work on a malaria vaccine. The efficient Japanese, for example, who began their war a little earlier than everyone else, had 151,000 cases with 10,000 deaths during their 1938 Wu Han campaign in the Yangtze River region.

armies, the colonialists, and even the natives, needed the shot of a malaria vaccine.

The malaria vaccine makers of the first half of our century tried to follow the precepts and techniques that were successful in making protective bacterial vaccines. They were soon to learn how simplistic and misleading the analogy was. The bacteriologists could grow their specific bacterium in a relatively simple medium and obtain, in washed, pure preparation, enormous numbers of the microbes. The cultured bacteria could then be inactivated for vaccine manufacture by such uncomplicated methods as adding dilute solutions of phenol or formol. The vaccine maker could usually test the efficacy and potency of his vaccine in a variety of experimental animals—mice, guinea pigs, monkeys—who were susceptible to infection. These animals could be given a series of immunizing injections with the vaccine under trial and observed for any untoward effects that might preclude its use in humans. Blood from these animals could be obtained and the serum antibody elicited by the vaccine measured to give a reliable test tube assay of protective potency. Finally, the experimental animals could be challenged with an injection of living, "hot" bacteria to determine just how immunizing the vaccine really was. Escalation from the experimental animal to the experimental human was reasonably straightforward and uncomplicated for the early bacteriologist-immunologists, who had no human and experimental animal experimentation committees or national regulatory agencies to satisfy.

Then there was the malaria parasite—parasites, really, since so many species infect almost the entire kingdom of vertebrates from snakes to humans. None of those species could be grown in culture. If the researcher wanted material to make a vaccine from the parasite's blood stages, it was necessary to "grow it up" in an infected bird, or monkey, or human. If the researcher wanted to make a vaccine from the mosquito stage, it was necessary to "grow it up" in a batch of colonized mosquitoes. Nor were there any experimental animals to test a vaccine made from a human malaria parasite; those parasites could only infect one kind of experimental animal—the human. It was, and to a great extent still is, all very host-restrictive as far as malaria vaccines were concerned. One could, for example, make a vaccine from a bird malaria parasite and protect a chicken. Then one would have to argue, "Hey, I've got a great vaccine for chickens and it should work on humans equally as well." That would be a very big assumption. For one thing, the bird malaria vaccine (or rodent malaria, or monkey malaria) would be so different (antigenically dissimilar) from the human parasites that it would not protect humans against *their* malarias. For another thing, even if you made a vaccine of human malaria parasites in the style of the chicken vaccine, there would be no guarantee of similar success.

Another malarial complication is its stage-specific diversity. A bacillus is a bacillus is a bacillus; but a malaria parasite is successively a sporozoite, exoerythrocytic schizont, merozoite, trophozoite, erythrocytic schizont, gametocyte (male and female), gamete (male and female), ookinete, oocyst, and back to sporozoite. All stages are different; each would require its own vaccine, or there would have to be a vaccine composed of material derived from several stages. For example, if a vaccine were to be made of merozoites (the stage that invades the red blood cell), then it would not protect against the invasion of the liver cell by the sporozoites (the stage in the mosquito's salivary glands that invades the human liver cells). Conversely, a vaccine made of sporozoites would not protect against the merozoites. Thus, any sporozoite vaccine would have to induce a very solid immunity; if any sporozoites did escape to establish an infection in the liver, then the disease from the ensuing invasion of the blood would be, for all practical purposes, as intense as if no vaccine at all had been given. And for another example: if a vaccine were prepared from the gametocytes or gametes, then possibly transmission through the mosquito might be aborted and the prevalence of malaria infection gradually reduced. Meanwhile the clinical course of the infected individual would remain unaffected. Thus, a gametocyte vaccine would have to be accepted altruistically by a population, or else it would have to be given under another guise (i.e., you lie to them), or you would have to give mass antimalarial chemotherapy coincidentally with the immunization campaign. Not impossible but difficult—very difficult (assuming of course that there *was* an effective gametocyte vaccine; and that's a further very big assumption).

Another problem. Malariologists had long noted the strain-specific nature of naturally acquired immunity to malaria. If an adult born and raised in a locality of intense perennial malaria transmission *(holoendemicity)*, having survived childhood malaria and having become immune, moved to another endemic area, he would come down with a clinical attack of a new malaria infection. His immunity would be good only for the malaria strains in his immediate vicinity. A strain of malaria parasite from another area, sometimes just a few miles away, may be so different that there is little or no cross-protection. One must also keep in mind that there are four species of malaria parasites, all antigenically different, all having strain variation (although target vaccines would be for the two most important parasites, *Plasmodium falciparum* and *Plasmodium vivax*).

And another problem. Even a protective vaccine, to be of practical value, would have to confer a relatively long-lasting protection. For soldiers, tourists, and business people who would spend a short period of time in a malarious locale, a short-acting vaccine would be useful. It could also be an expensive vaccine because those users could afford it. However, for the truly malarious, the impoverished villagers of the Third World, the vaccine would not only have to be cheap, it would also have to give an "extended warranty." Rounding up a population for mass vaccination every six months, or every year, or even every two years would be near impossible. Today, the World Health Organization's Extended Program of Immunization campaign to immunize Third World children with proven, highly effective vaccines such as DPT (diphtheria-pertussis-tetanus) may capture, at best, 30 percent of the target population. And when you look *very* closely at the data, you will often find that the majority of that 30 percent are the children of middle-class city dwellers. Also, the nature of malaria transmission is such that vaccination coverage and the immunity it afforded would have to be even better than for existing bacterial and viral vaccines. Epidemiologists have calculated that every existing case of smallpox has the potential to infect four or five other people; but in malaria, with an efficient vector such as the African *Anopheles gambiae*, each case (gametocyte carrier) has the potential to infect one hundred other people.

These difficulties were, and to a great extent still are, formidable obstacles in the making of an effective malaria vaccine. Nevertheless, the malaria vaccine researchers of the first half of this century persisted and, *nil desperandum,* tested vaccines made in the manner of bacterial vaccines. The vaccines used in most of these trials were made from either parasite-infected red blood cells or plasmodial material extracted from infected blood and treated-inactivated with a dilute solution of formalin. In this way, the first vaccination experiments employed avian and primate malarias. The birds and monkeys were given a series of immunizing injections with the formalinized vaccine, and this was followed, sometime later, with a challenging inoculum of living parasites. Nothing much happened in any of these trials; the vaccines afforded the experimental animals—birds and beasts—little or no protection. It was all quite discouraging but again, *nil desperandum*, maybe things would go better with humans.

America's entry into World War II and the intractable problems of military malaria gave renewed impetus to the search for a malaria vaccine. A team was established at the Rockefeller Institute (later Rockefeller University), led by Dr. Michael Heidelberger. They went straight to the experimental animals who were inexpensive and in plentiful supply—the paretics and the prisoners.

Heidelberger and his group were confined to *Plasmodium vivax* because it could not cause any serious or fatal infections and also, for the same reason, was the parasite used to treat neurosyphilis. A vaccine was made from a formalin-inactivated extract of parasites and injected into paretics and prisoner volunteers in what would now be considered a heroic and unacceptable immunizing regimen with an unacceptable vaccine. The recipients were given daily injections for four to seven days, and the course repeated one or more times at two- to three-week intervals. These injections with a vaccine that was "rubbishy" with bits and pieces of foreign blood matter were given subcutaneously, intramuscularly, and intravenously. The

FDA would now be horrified if we used such a vaccine in guinea pigs, let alone humans. Yet impure and imperfect as it was, the vaccine was reported to have caused little or no untoward reactions or discomfort in the recipients. On the other hand, it didn't cause any immunity, either. When the vaccinated were challenged with either *Plasmodium vivax*-infected blood or the bites of infected mosquitoes, there was no evidence of immune protection; they developed malaria with the same frequency and to the same intensity as the unvaccinated control group of humans.

Well, reasoned Heidelberger & Co., if the vaccine hasn't got the moxie to induce protective immunity in those previously unexposed to malaria, the "immunological virgins," maybe it will be strong enough to tip the balance toward protective immunity in those already infected with *Plasmodium vivax*. Vivax (benign tertian) malaria is characterized by a series of relapses, emanating from the parasites persisting in the liver, for up to about five years. Would vaccination either prevent or abort these relapses? Two hundred "prisoners" with a known history of vivax malaria were collected and given a series of two or three injections of a vaccine consisting of an estimated 2 to 3 billion dead *Plasmodium vivax*. Again, no immunity; those given the vaccine relapsed at the same rate and way as the unvaccinated. Reading the 1946 publication reporting the results of this experiment, one wonders where all these previously infected "prisoners" came from; the description of the human subjects in the paper is very vague. Paul Silverman, who in this narrative will presently play an important role in post-World War II malaria vaccine research, says that he has seen the original records and that none of the two hundred prisoners were American felon volunteers but were, in fact, Italian and German prisoners of war who had become infected while fighting the Allies on malarious battlefields.

Both experimental animals and experimental humans, it seemed, simply did not possess the immunological constitution to mount a protective response from "conventional" malaria vaccines. At the conclusion of the disappointing human experiments of the 1940s, the researchers were faced with unresolved questions: Could a more potent vaccine ever be developed? Or was there something about the malaria parasite that prevented the host from reacting effectively to a vaccine in a way that would confer solid immunity or even a partial immunity for "clinical" protection? Any host, any vaccine.

Jules Freund at the Division of Applied Immunology of the Public Health Research Institute of the City of New York found that the vaccine *could* be improved by the addition of a non-specific stimulating agent called an adjuvant. This adjuvant, still known as Freund's adjuvant, is an alchemist's cocktail of a non-pathogenic bacterium related to the tuberculosis organism *(Mycobacterium)* suspended in an emulsion of mineral oil and water. From 1945 to 1948, Freund and his colleagues undertook a series of immunizing experiments in animals vaccinated with the parasite-adjuvant mixture. They began with ducks.

Plasmodium lophurae came to the Bronx from Borneo. It was found as a harmless infection in a captive, beautiful fireback pheasant in the New York Zoological Gardens. *Plasmodium lophurae* may be harmless to fireback pheasants, but it is death to ducklings. Freund found that the adjuvant-lophurae vaccine, if given in three injections one month apart, would solidly immunize some birds, while in the others only a low-grade, self-curing infection would develop. But there was a price; the adjuvant caused a fatty degeneration of the liver. From the ducklings they went to monkeys to test a *Plasmodium knowlesi*-adjuvant vaccine in the highly susceptible rhesus monkeys. The vaccine was as effective in the rhesus as it had been in the ducklings, but again there was a price. The adjuvant often caused terrible lesions at the injection site, and later it was shown that it could induce auto-immune damage to the nervous system and other tissues. The adjuvant made vaccination *theoretically* possible, but it could never be used in humans. It was too harmful even to be *tested* in humans. Thus, the mid-century brought research on the malaria vaccine to a virtual dead-end halt. However, the urgency was gone. The war was over, leaving only the natives to their malaria, and even for that huge segment of humanity there were the quick fixes available from the new miracle chemicals: DDT and chloroquine. Chemistry, and not immunity, was about to free the world of malaria—forever!

Forever lasted until about 1965. It had been a roller-coaster decade for malaria, beginning with the 1955 World Health Organization Assembly's endorsement of the global eradication plan. However, by the mid-1960s, the reality of malaria-as-always in Africa and explosive new outbreaks in Asia prompted the statement by the late Leonard Bruce-Chwatt, a leader of the World Health Organization program (and a great gentleman), to write: "The wave of endemic malaria that has swept over southern Asia may well undo all the achievement of the last two decades." The near-total collapse of malaria eradication programs throughout the tropical world was of great embarrassment to the planning and funding agencies, who were forced to retreat to their own new lines of self-justification—and self-perpetuation. The United States Agency for International Development was outstandingly embarrassed.

AID is the agency that gives our money away to foreigners. Its health activities have always been something of a joke to the experienced professionals, but the Agency did little ostensible harm; it waved the American flag and, presumably, captured the minds and affections of tropical peoples and their governments. AID always liked the Big Project, and since there was nothing bigger at the time than the Global Eradication of Malaria Program, it was a natural for largesse. From 1955 to about 1970, AID contributed approximately $1 billion to the World Health Organization and various national malaria eradication programs. All that taxpayer money down the drain. AID was now looking for a quick fix. Reenter the malaria vaccine as sold by Paul Silverman.

Chapter 16

The Selling of the Malaria Vaccine

WHEN I WAS a graduate student at the University of London's School of Tropical Medicine and Hygiene shortly after World War II, I heard about, but never met, the "other" American at the "other" School. Paul Silverman was then at the Liverpool School of Tropical Medicine, pursuing his doctoral research on the biology of tapeworms. As a new Ph.D., politics and science took him to Israel to work on leishmaniasis under Saul Adler. A few years later, Silverman returned to England, where he went to work as a research parasitologist for the pharmaceutical firm Allen Hanbury's. There he developed and patented vaccines that claimed to give partial protection against certain parasitic worm infections of sheep, calves, and horses. In 1963 or 1964, Silverman recycled back to the United States to join the University of Illinois biological sciences faculty. His experience with the worm vaccines led him to the opinion that it was possible to immunize against other parasitic infections, including malaria. The great majority of parasitologists and immunologists at that time did not agree with this view. When Silverman got a million-dollar grant from AID to develop a malaria vaccine, it made the doubters into instantaneous believers.

In 1965, the World Health Organization invited a group of people to its headquarters in Geneva to suggest any innovative ways that the failing Global Malaria Eradication Program might yet be rescued. Although Silverman never had any "hands-on" experience in malaria research, he attended the meeting to advance his ideas on the feasibility of a malaria vaccine. However, one non-scientist participant, Lee Howard, who was head of the health division of AID, did listen to Silverman's arguments and was intrigued. Their flight bookings were rearranged so that they could return together; over the Atlantic, Silverman, a persuasive man, expounded on how he thought a malaria vaccine should be formulated.

Silverman held that an effective vaccine should be composed of antigens from two stages of the malaria parasite: the sporozoite antigen to prevent infection from the mosquito bite, and an asexual (blood-stage) antigen to induce protective immunity against any breakthroughs. That would seem to be a logical approach even by the criteria of current knowledge. But in 1965 there were enormous, if not insurmountable, technical hurdles to be overcome. In 1965, there were no techniques to culture the human malaria parasite (or any other malaria parasite) as a source of antigen for experimental vaccine testing. In 1965, there was no known way in which primates could be experimentally infected with human malaria parasites. In 1965, there was no known practical way to produce large amounts of pure sporozoites from either the mosquito or culture for use in vaccine trials. Silverman argued that these were technical obstacles that could be solved by a major, well-funded research program. Those were the key words, "major" and "well-funded": lots of people, monkeys, and money.

Several weeks later, Silverman was invited to Washington to make a presentation to the AID health division. He again made his case for research on a malaria vaccine. AID had no scientists and no malaria researchers of its own. AID had been solely in the "handout" business—food, insecticides, advice, etc. AID had never conducted or supported basic research. It had no scientist administrators, no scientist consultants for peer review or evaluation. No nothing except, perhaps, an accounting system that would allow the staff to divert funds to any project they wished. AID did send Silverman's proposal to an ad hoc panel of respected malaria researchers. The panel members were unanimous in their judgment that it was not feasible to develop a practical malaria vaccine for humans; they were also of the opinion that AID should stay out of the research business. AID then took the first step in establishing a policy pattern that was to continue for the next twenty-five years—it ignored the advice of its expert consultants. Silverman was given $1 million to develop a malaria vaccine. Even Silverman was surprised by the amount of money he was being offered.

Ed Smith became AID's project officer—and this was the beginning of yet another peculiar policy. Smith, an employee of AID, was a medical entomologist. He knew little about immunology or about the epidemiological or clinical aspects of malaria. Later, Smith was replaced by James Erickson, who wasn't even a medical entomologist; his entire professional career had been concerned with agricultural pests. These men answered only to higher functionaires as devoid of the needed expertise as themselves. Yet, Smith and Erickson ran the project in an authoritarian fashion. They took expert advice when it suited them and rejected the advice that did not suit them. Those deemed to be critics were excluded from making site-visit evaluations to judge the merit of ongoing research.[67]

With these protective arrangements in place, Silverman and his group began work in 1965. The objective, of course, was to immunize humans, not monkeys, but it was neither practical nor possible (and certainly not ethical) to carry out human trials with an unproven vaccine. Silverman had his designs on the prisoners of the Statesville Prison at Joliet, Illinois, where antimalarial drug studies had been carried out. However, for the moment he had to be satisfied with *Plasmodium knowlesi* in rhesus monkeys. Infections with this parasite are so consistently lethal for the rhesus that it was felt that if you could devise a vaccine to immunize this highly susceptible monkey, then there should be no difficulty in adapting the technique to protect humans against *their* malarias.

Over the years a lot of monkeys have been sacrificed to this proposition, although it has never been proven that there is an experimental extrapolation, in respect of a malaria vaccine, from monkey to man. Not only was there the starting uncertainty that any of the results from monkeys would be meaningful, there was the other difficulty, which became evident not long after beginning the project, that a basic block of the research would have to be dropped. Having failed to devise a practical method for sporozoite antigen production, it would be impossible to make the polyvalent vaccine composed of a mixture of sporozoite and blood-stage antigens. There was a bitter disagreement with the group given the subcontract to work on the sporozoite antigen.

67. A 1989 General Accounting Office (GAO) report of their investigation of the AID malaria vaccine program noted that there was "Inadequate oversight and review of the project" and that the "Criteria for selecting onsite review team members [was] questionable."

The evaluation was that they were not working at a "certifiable level of effort." Even the generally permissive AID considered the research value so feeble that it withheld payment to the University of Illinois. Although this line of research ended in 1972, the university did not get its money until 1975.

Five years and about $1.5 million later, Silverman's group had done the bishop's work. They had confirmed Freund's finding of the 1940s and those of the English workers of the early 1960s that indeed the rhesus monkey could be immunized with a *Plasmodium knowlesi* vaccine.[68] However, the old problems remained. The vaccine would induce protective immunity *only* when it was used with Freund's adjuvant. Therefore, it was impossible to proceed to testing in humans. The level of protection provided by the antigen-adjuvant vaccine was good for a monkey, but in humans a *Plasmodium falciparum* infection of similar intensity (2 to 5 percent of the red cells parasitized) would cause acute symptoms and pathology. Then there was the question of "purity." It is one thing to inject a monkey with foreign red blood cells; but despite the time-proven practice of blood transfusion, a malaria vaccine composed of whole infected red blood cells or parasite preparations contaminated with red blood cell fragments was deemed unacceptable for human use. Eventually, the "purists" would win the day, and the AID contracts, but in 1970 the biotechnology to make a pristine parasite antigen was not yet available.

Silverman had hired an osteopathic physician, Lawrence D'Antonio, to address the purity problem. D'Antonio, who had previous experience in malaria research at the

68. These were studies carried out at the National Institute for Medical Research in London by Geoff Targett and John Fulton in 1965 and by K. N. Brown and his colleagues in 1970.

Walter Reed Army Institute of Research, used a French pressure cell to winkle the parasite out of the red blood cell. By this method, the parasitized blood was placed in the apparatus, the pressure pumped up, and a needle valve opened to allow emission of the blood. When the internal pressure within French cell and the flow through the valve were properly adjusted, the rapid decompression would rupture the red blood cell membrane without harming the parasites. The freed, intact malaria parasites were then collected, washed up by repeated centrifugation, inactivated by heat or formolization, and used as an immunizing vaccine. Electron microscopy showed that there was still some adherent red blood cell membrane material about the parasites, but it was a "cleaner" preparation than anyone else had obtained.

It was a neat technique. However, when Silverman was about to go to a meeting at which he was to read a paper describing the vaccine preparation by the French pressure cell method, he received a phone call from the AID office in Washington informing him that he couldn't present his paper. D'Antonio had filed for a patent to cover the method, and for the moment it was no longer government property, even though the research and D'Antonio's salary had been paid for by the taxpayer. Meanwhile D'Antonio had left Urbana ("disappeared" Silverman says), to resurface some six months later at a college of osteopathic medicine in Philadelphia. Eventually the French pressure cell technique was returned to government ownership; but D'Antonio's action began a series of deals in which malaria vaccine research was to become a business.

In 1972, Silverman moved his show to the University of New Mexico in Albuquerque, where he began looking for alternatives to Freund's adjuvant. Freund's adjuvant, to refresh your memory, is an emulsified suspension of a nonpathogenic *Mycobacterium* species in mineral oil. The mineral oil component is largely responsible for the adjuvant's toxicity since it is metabolically inert. It is not metabolized by the body and tends to persist at the injection site, where it causes inflammation that frequently progresses to abscess formation. Silverman replaced the mineral oil with peanut oil, which is biodegradable by the body. The *Mycobacterium* of the Freund's adjuvant was replaced with the more acceptable BCG *Mycobacterium,* which was being used in humans to vaccinate against tuberculosis.

The new adjuvant didn't work quite as well as did Freund's, but it did work. Thus, in 1973, Silverman thought he had a technique to make a vaccine "clean" enough and an adjuvant safe enough to take that giant step to try to immunize humans against malaria. Since there was no known method to maintain malaria parasites in continuous, long-term culture, the scheme was to go to a place where *Plasmodium falciparum* malaria was prevalent, find a patient with an acute infection, and make a vaccine (monkey-fashion) from his or her blood. That vaccine would then be used to immunize a group of volunteers. It all had to be done in a Third World malarious country that would participate in the whole investigation. The FDA would never approve of the experimental vaccine for use in Americans or anybody else. In 1973, Silverman got his collaborating country—Brazil.

A medical research institute in Manaus agreed to collaborate. They would find the patients infected with malaria parasites as the source for the vaccine and then conduct the immunizing trials in volunteers. Had the investigation actually come off, it might have changed the course of malaria vaccine research. Success or failure in the definitive human host might have injected a more substantial element of reality into the AID project. However, Silverman's vaccine never came to trial. In 1973, the Arabs imposed their oil embargo upon the entire dependent world. Brazil's marginal economy was devastated by its inability to buy cheap fuel. In looking for the least painful and least politically harmful budget cuts, the research institutions (as usual) were the first to get the ax. The institute at Manaus was closed, and for the American researchers in New Mexico it was back to the monkeys. Other scientists in other countries now began to enter the race for the malaria vaccine. Science is like that; it is not exactly the same as the feeding frenzy of sharks, but a major, lavishly funded research program does make other scientists take notice. Nobody wants to be left behind. The English scientists especially didn't want the vaccine to go to the Americans by default. English tropical medicine research had lost a good deal of its stamina since the Empire had been disassembled, but this was still the nation of Ross and Manson. The tradition, interest, and expertise in tropical medicine lived on. Peculiarly, the vaccine research was not taken up by either of England's famous schools of tropical medicine in London and Liverpool, but by Sidney Cohen, a professor of pathology at Guy's Hospital and Medical School in London. Cohen was also confined to using *Plasmodium knowlesi* in the rhesus monkey and a vaccine with Freund's adjuvant. However, he took the purity principle one step further by making a vaccine entirely of the merozoite stage of the parasite.[69]

69. The asexual division of the parasite within the red blood cell (and liver cell in pre-erythrocytic schizogony) is completed with formation of the merozoites. It is these minute, 2 / 25,000th of an inch-long forms

By a rather painstaking process, Cohen's group were able to collect enough pure merozoites to make a vaccine to immunize a few monkeys. It was not known how one could get enough *Plasmodium falciparum* merozoites to vaccinate 200 million people, but it was felt that if the monkey preparation worked, technology would follow discovery. In their published reports, the Cohen group claimed that the merozoite vaccine afforded as good as, if not better, protection as the Silverman "semi-pure" vaccine. Even so, it still needed the toxic Freund's adjuvant to induce immunity, and it still did not give the solid enough immune protection that would be necessary in humans. Nevertheless, the general belief was that progress was being made in the development of a malaria vaccine for human use. The immediate question was, who had the better vaccine, the Americans or the British? What they did to resolve the rival claims was extraordinary. They staged a vaccination contest: the First (and last) New Mexican Invitational Malaria Vaccine Shootout.

To approximate reality, both groups would use a vaccine that had been prepared before the beginning of the contest. After all, making an extract or isolation of a malaria parasite for vaccine use isn't quite the same as making fresh orange juice. The vaccine must also have a longer "shelf life" than the orange juice because if it is to be applicable for field use, it must be relatively stable. Within the past decade, the global campaign of mass childhood immunization has encountered great difficulty in the Third World because of the frequent inability to maintain an uninter-

that are released into the bloodstream to invade new red blood cells and perpetuate the cycle. The merozoites are "pure" in the sense that they are the only "blood" stage of the parasite that is entirely free of host cell membrane or any other host-derived material.

rupted "cold chain" from the manufacturer to the village.[70]

Cohen's delegates brought their vaccine, carefully preserved under refrigerated conditions, to Albuquerque. Silverman's group had their freeze-dried vaccine. Rhesus monkeys were vaccinated with each vaccine by the respective preferred immunizing schedule. Some monkeys were left unvaccinated to serve as "yardstick" controls. When the monkeys were deemed to have been immunized, they were "challenged," along with the control animals, with an injection of infected blood. The controls died of a fulminating infection as expected. Half of Silverman's immunized monkeys and all of the Cohen monkeys died. Not exactly encouraging results after seven years and several millions of dollars. Although both vaccines gave disappointing results, the British were particularly embarrassed. Their vaccine had no immunizing potency whatsoever. In defense, they claimed that their vaccine did not "travel well." That may be true, said the Silverman people, but if you can't ship a vaccine from England to America under the best possible conditions, how are you going to get it from England to a remote village in Africa where, ultimately, a vaccine would have to be given? Nevertheless, the British were given a "Mulligan" and were allowed to immunize a new batch of rhesus with a new batch of vaccine, which they shipped from London. This time only 50 percent of the immunized monkeys died. Silverman also blamed the poor showing of his vaccine on storage problems and said a freshly prepared vaccine had greater immunizing potency.

Despite the results of the competition, the American research establishment (mostly Eastern division) did not rally to Silverman's side. They felt that although the Cohen vac-

70. There is a fuller discussion of the "cold chain" problem in chapter 13 of my book *The Thorn in the Starfish*.

cine had proved to be impotent, it was the better vaccine. It was purer. It was the way to go. Besides, the underlying feeling was that Cohen was a more acceptable, less controversial scientist than Silverman.

Silverman now speaks of that time with great bitterness. He claims that not only was he not supported by his peers but they also vilified him by spreading rumors questioning his honesty, his sexual orientation, his marital fidelity. "Nonsense," say those peers, their criticism was only that they thought his science was poor (except, we might note, that the peers who inherited Silverman's objective of a malaria vaccine, and his AID funding, have not been able substantially to improve upon the original results). Silverman now began to drift out of the project, having found a new forte—administration. He became Vice President for Research at the University of New Mexico and, a few years later, Chancellor of the entire State of New York University system. Not a bad career accomplishment for that "other American" from that "other" school in England. And to his credit, it was he who initiated the idea that it was appropriate to return to research on a malaria vaccine.

After Silverman departed the scene, Karl Rieckmann took over the project. Rieckmann was a German physician who went to Australian New Guinea after World War II, where he was employed in the malaria control program. He then emigrated to the United States, where he was taken on by Silverman, who hoped that he would help conduct the human immunization experiments (which never materialized). For a few more years the project continued in New Mexico under Rieckmann, until in 1978 or 1979 it was rather abruptly phased out.

One would think that after fifteen years of supporting highly funded research that had essentially gone nowhere—

it was no closer to a vaccine to immunize humans against malaria than Freund and Heidelberger had been in 1945—AID would call it quits. If you think that, then you are unschooled in the ways of government bureaucracy. In 1980, AID did what any compulsive gambler would do; it doubled the stakes and spread its bets.

AID doubled the research funding to a level "which will permit completion of this important project." But now instead of the one winner-take-all policy, it became essentially a three-winner contest. The winners were Ruth Nussenzweig, who had come to New York University from Brazil; Wasim Siddiqui, who had come to the University of Hawaii from India; and Miodrag Ristic, who had come to the University of Illinois from Yugoslavia.

Chapter 17

The Sporozoite Follies

RUTH NUSSENZWEIG is a brilliant scientist who has built and maintained a research empire in New York. She and her immunologist husband, Victor, took on the difficult project of making a sporozoite vaccine as their piece of the AID-funded action. I noted earlier that there are formidable problems associated with making a vaccine from sporozoites. There is the problem of isolating and purifying the sporozoites from the mosquito and there is the problem that any immunity must be absolutely solid. Because malaria immunity is stage-specific, a single sporozoite that escapes can result in an infection as intense as if there were no immunity at all. Despite these difficulties, there was in fact as much, if not more, promise from experimental precedence for the Nussenzweigs than for those engaged in trying to make a vaccine from other stages of the malaria parasite.

In 1941 and 1942, an international research team consisting of Paul Russell, the American, Hugh Mulligan, the British officer of the Indian Medical Service, and Badri Nath Mohan, the Indian scientist, working at the Pasteur Institute of Southern India in Coonoor, had carried out pioneering experiments on sporozoite immunity. Employing an avian malaria parasite, *Plasmodium gallinaceum*, which often produces fatal infections in chickens, they showed that fowl could be immunized by injections of sporozoites that had been exposed to an ultraviolet light source. In this way, the sporozoites were inactivated; that is, they were still alive and capable of exciting an immune response but no longer able to invade the host cells to produce an infection. The irradiated sporozoite vaccine was able to protect chickens from the bite of mosquitoes carrying *Plasmodium gallinaceum* sporozoites. It required a long series of immunizing injections to do this, and even so only about half the immunized chickens were solidly protected. They also showed that the immunity was stage-specific. Birds that were completely protected from the bite of the mosquitoes became infected when inoculated with parasitized blood. There was no claim that sporozoite immunization would be a practical, realistic way to protect humans. However, it did work in some chickens, and it was a start.

Twenty years later, in 1968, the New York University researchers found that if the sporozoites were inactivated with a carefully adjusted dose of X-ray, they would induce a much higher level of immunity than sporozoites inactivated by ultraviolet light. With these X-ray-inactivated sporozoites, they were able to produce a solid immunity in rodents against rodent malaria and then, as a crowning experimental achievement, to immunize monkeys against a primate malaria. The results were so good in the laboratory animals that human trials were begun by 1971.

One crucial study, paid for by the U.S. Army, was carried out by researchers from the University of Maryland and New York University. David Clyde of Maryland was in charge. Clyde had long experience in clinical malariology research in the tropics. He was, like Ross, a physician who had been born in India and whose father was a general in the Army of the Raj. He had been conducting experimental malaria studies since coming to the University of Maryland

and had access to prisoner volunteers at the state jail in Jessup. The immunization trial that Clyde organized was so complicated that it could be carried out only under the most controlled conditions.

First, volunteers had to be infected with *Plasmodium falciparum* to serve as a source of gametocytes. Unfortunately, gametocytes appear only after several bouts of intense fever during the course of this potentially fatal infection. Nor does every infection produce gametocytes. By the time the study was completed, thirty-three men had to be infected to yield six infections with gametocytes. Next, *Anopheles* mosquitoes that had been bred in the insectory were fed on the gametocyte carrier volunteers.[71] The mosquitoes then had to be maintained under optimal conditions for about twelve days, until the "babies" had been made—the sporozoites in the salivary glands, at which time they were given the inactivating dose of X-ray.

About an hour after being zapped with X-rays, the mosquitoes were placed on the forearm of three volunteers. It required about one hundred mosquitoes per person, to feed and inject their inactivated sporozoites into the bloodstream. This whole business had to be repeated four times to get what they thought to be an adequate immunizing experience. Two weeks after the last "experience," the volunteers were challenged with "hot" mosquitoes carrying non-irradiated sporozoites. The results: Two of the three volunteers came down with malaria. No protection whatsoever. One volunteer seemed to be immunized and did not become infected. The conclusions? Despite a mere 33

71. The study was ethical in that any volunteer who seemed to be in trouble was immediately treated. However, for reasons that were not explained, multi-drug-resistant strains of *Plasmodium falciparum* were used. One strain had a reduced sensitivity to the drug of last resort—quinine.

percent effectiveness and despite its being, for all practical purposes, an impossible technical procedure, Clyde et al. celebrated the results and called it "an encouraging step toward the goal of immunizing man against malaria."

About the same time that Clyde's group were carrying out their human immunization trials in Maryland, the U.S. Navy, not to be upstaged by the other service, was carrying out a similar study at the Statesville Prison in Illinois. Their results were not any more encouraging than those from the Maryland prisoners. In three volunteers who were immunized with X-ray-inactivated sporozoites, all were protected against challenging mosquito bites for a few weeks. All three came down with malaria when similarly challenged a few months later. This indicated that at best the protection afforded by vaccination with irradiated sporozoites was of short duration. The group's conclusions from these results? ". . . Protection in man can be induced by the inoculation of irradiated sporozoites," and ". . . the successful immunization of human volunteers reported here should encourage continuing efforts toward the development of a sporozoite vaccine against human malaria."

When he was student at the London School of Hygiene and Tropical Medicine, his friends called him "Golden Boy." Amour propre notwithstanding, Bill Bray, the Golden Boy, became a good scientist who viewed the "great events" of medical parasitology with critical discernment. In 1975, Bray was director of Britain's Medical Research Council laboratories in Fajara, the Gambia, a torridly humid ribbon of a West African nation whose landmass consists of a few inland miles along each bank of the Gambian River.

Malaria is an everyday matter of life and death in the Gambia. Children are particularly vulnerable to its most acute, frequently fatal, form. The Medical Research Council laboratory was, as it still is, deeply engrossed in malaria investigations. Thus, the American irradiated-sporozoite vaccination trials in prisoners were certainly an "event" to be followed and repeated, even though Bray and some other malariologists read the author's results as two-thirds failure rather than the self-proclaimed one-third success.

The question Bray wanted to answer was whether semi-immune African children would be better protected by the vaccine than the immunologically "naive" prisoner volunteers had been. If it was more protective, if it reinforced the slowly awakening childhood immunity under natural endemic conditions, then the malaria vaccine would truly be in business despite concerns over the technical and administrative difficulties in vaccinating native populations. If it was less effective, then forget it—or put the research money eggs in another basket. With great difficulty and nerve, Bray carried out the experiment that nearly ruined his career.

In bare outline it seemed to be a simple study. Get mosquitoes; infect them; irradiate them; feed them on volunteers; then send the volunteers back to their village, where they would be monitored for reininfection. Old Africa hands know the real heart of the matter, the great difficulty in carrying out a study of this sort. Bray bred *Anopheles gambiae* mosquitoes, marvelously efficient carriers of *Plasmodium falciparum,* in his laboratory's insectary. They were then fed on Gambians who were found to have *Plasmodium falciparum* gametocytes in their blood. Since there was no irradiation source in the Gambia, it was necessary to take the mosquitoes to Accra, where there was a Cobalt 60 irradiation apparatus at the Ghana Atomic Energy Commission. In 1976, Ghana had an Atomic Energy Commission and an impoverished citizenry.

Bray and his mosquitoes made the journey despite the customary African travel hazards of canceled flights and confirmed air bookings turning unconfirmed because someone gave a bigger "dash" to the airline agent. With a suitable "dash" to the Customs inspector, Bray and the now irradiated anophelines rapidly made their way back to the Gambia, where he had two six-year-old children lined up to make medical history. Each child, along with two age-matched control children, had been given a course of chloroquine to clear any and all parasites they might have had. The irradiated mosquitoes were quickly taken from the airport to the laboratory, where they were dissected to remove the sporozoites from their salivary glands. Approximately 500,000 irradiated sporozoites were inoculated into the arm of each of the two children. Two weeks later the whole procedure—chloroquine treatment followed by vaccination with irradiated sporozoites—was repeated and the four children (two vaccinees and two controls) returned to their village, where they were again assaulted nightly by malaria-carrying anophelines. Three weeks later, all four children, the two controls *and* the two vaccinated kids, had malaria. The irradiated-sporozoite vaccine did not work in the endemic setting for which it had been intended.

If the vaccine had been effective, Bray would have been a hero. It didn't work, and sensibilities were offended. The results were ignored and instead, Bray the Ph.D. was accused of playing Bray the M.D., a professional qualification he did not hold (although he now maintains that he had the approval of his human experimentation committee). Shortly thereafter he lost his job as director and was given such assignments as Baghdad and Addis Ababa. It wasn't until much later in his life that he was allowed to return to England, where he was placed in the Research

Council's satellite laboratory at the Imperial College of Science and Technology.

These human experiments confirmed that man was not a mouse. Later, more sophisticated research revealed that the human immune system "turned a blind eye" on certain important components of the vaccine that the mouse "saw." Furthermore, even if immunologically "seen" in an appropriate way, those components were highly variable. Thus by 1980, the millions of dollars AID had invested in the malaria vaccine project had yielded nothing of practical value. However, the Nussenzweigs were moving with the biotechnical times, and through a remarkable piece of research they saved their project and, temporarily, AID saved face.

The Nussenzweigs' premise was that biotechnology would provide the right vaccines. What had gone on before was "crude," "impure." Immunogenetic gene cloning, recombinant technology, monoclonal antibodies, peptide synthesis—these would yield the breakthrough to the sporozoite vaccine.

The Nussenzweigs, as had others before them, observed that when sporozoites were placed in serum containing antibody, a coating formed about them and they became immobile (the sporozoites, not the Nussenzweigs). The important immune reaction that would disarm them, make them non-infective, took place on the "skin" of the sporozoites. It was very specific to the proteins (antigens) on that "skin" and the antibody that the host had elicited to them. Each of those protein antigens would be manufactured under the instructions of a specific gene. By 1980, genetic engineering technology had so advanced that it became possible to isolate a gene, insert it into a bacterium or yeast, and then those "combined" microorganisms would genetically "think" that they were a mouse, or a human—or a sporozoite. In this way, the New York University group was able to get a bacterium growing in a culture tube to manufacture the "skin" protein (circumsporozoite protein) of the sporozoite. Exit the mosquito. Enter the culture tube. The sporozoite could be now manufactured in "pure" form and, theoretically, unlimited quantities could be obtained.

Even more stunning was the research that led to the synthetic manufacture of the circumsporozoite antigen. When the circumsporozoite antigen was chemically analyzed, it was found that the active antigenic "piece" was a simple quartet of four amino acids repeated over and over again. With the new peptide synthesizing machines, those four amino acids could be taken off the laboratory shelf, put into the machine, the machine programmed—and out would come the circumsporozoite malaria vaccine. It was as easy as making Saran Wrap. Exit the culture tube. Enter the synthesizer.

The totally synthetic vaccine seemed to be the real thing. Antibody from mice that had been vaccinated with it combined strongly with living sporozoites obtained from infected mosquitoes. In fact, the mice reacted so well that it was a foregone conclusion that humans would react equally as well. That humans could finally be immunized against malaria with a synthetic, stable vaccine capable of being produced in bulk lots.

The Nussenzweigs and New York University were now prepared to put their vaccine to the definitive test in humans; but before doing so, there were some business matters to be concluded. NYU and the Nussenzweigs applied for a patent on the synthetic vaccine. All their research had been paid for with taxpayer dollars. And all those years NYU had been collecting approximately 65 percent off the top.[72] Then NYU and the Nussenzweigs entered into an agreement with Genentech, a commercial biotechnology firm. Only the World Health Organization, with a small financial stake in the research, protested in rightous indignation and declared that it would legally contest the patent. AID said, in effect, "Take it all, only give us the vaccine." AID now created a special policy which permitted its funded researchers full rights to their inventions and allowed them to collaborate with an industrial partner. It seemed to be a betrayal of the program's altruistic objectives. The comic aspect of this unseemly rush to profit was that the vaccine was still imaginary. Not a single human being had been vaccinated with it, let alone protected by it.

During the next two years there were two human trials with the new high-tech sporozoite vaccines involving nine volunteers. Of the nine vaccinated, only two were protected from acquiring malaria following the challenging bites of infected mosquitoes. Moreover, one of the two successfully immunized people developed an allergic reaction to the vaccine—sneezing, itching, watery eyes. The report makes light of this, but in a New Guinea village, for example, it certainly would not be a "something-nothing," as they say in New Guinea villages. If you went to the village and vaccinated fifty people and five got immediate allergic reactions, you could kiss your immunization campaign goodbye.

Incredibly, AID continued to be wildly enthusiastic and

72. There were also some alleged fiscal irregularities with NYU's administration of the malaria vaccine grant's funds. The congressional General Accounting Office investigators found "about $102,000 paid to New York University for excess salaries and related fringe benefits and overhead"; in addition, "auditors were unable to account for equipment costing about $52,000" (GAO report to the Honorable Daniel Inouye, U.S. Senate, October 1989, p. 34).

did mischief by making promissory statements such as, "Technically speaking a vaccine may be available for human testing as early as 1985," and, "AID could put the first new, major effective antimalaria weapon [the vaccine] since DDT to use in developing countries by 1990," and, "In August 1984, the AID announced a major breakthrough in the development of a vaccine against the most deadly form of malaria in human beings. The vaccine should be ready for use around the world, especially in developing countries, within five years." Hope sprang eternal even as late as 1986, when the director of AID's Bureau of Science and Technology announced: "We are on the verge of having a prototype vaccine for the sporozoite stage of malaria of Plasmodium falciparum." In another press release, AID announced that there "would be a [malaria] vaccine ready for widespread use by 1989." Beleagured health authorities in poor tropical nations heard these statements, believed them, and stopped their expensive antimalaria operations to await the vaccine.

Other scientists, not dependent on AID money for their research, looked at the Vaccine Emperor and saw that while he was not entirely naked he was certainly down to his jockey shorts. By 1988, critical papers were being published in scientific journals with titles such as "The Real Difficulties for Sporozoite Vaccine Development," and "Forlorn Hope for Malaria Vaccine."

Even as the sporozoite vaccine was sinking, there came the report from another study which signaled a new warning that the sporozoite vaccine might constitute an actual danger rather than a benefit. Michael Holingdale and Virgilo Do Rosario of the Biomedical Research Institute in Rockville, Maryland, discovered that if antibody was elicited to the vaccine at an insufficient level to be fully protective, then it had a paradoxical *enhancing* effect. The transmission of malaria was enhanced by the antibody. More sporozoites developed in the mosquito that imbibed antibody than in the mosquito that drank of common clean blood. The implication of this finding was that if an immunization program with a sporozoite vaccine was not completely protective, there was the real possibility that there would be even more malaria than before the immunization campaign.

Science, the prestigious "trade journal" of scientists, summed it all up in its issue of July 29, 1988, with the comment that "the [sporozoite] vaccine trials were disappointing." This made the Nussenzweigs unhappy, and they wrote a strong letter of protest to *Science* claiming that protecting one of three volunteers was "encouraging." Furthermore, they objected to their work even being evaluated in a forum such as *Science.* It was their opinion that the vaccine project should be "evaluated by peer review and not unsubstantiated commentaries in scientific journals." This was a sentiment that did not seem to recognize that their mouse and monkey research had been represented to the lay press as the human vaccine just around the corner.

Chapter 18

The Vaccine Felonies

IN THE LATTER YEARS, the AID Malaria Vaccine Project chronicles may seem, in good part, more appropriate to the *Police Gazette* than within the dedicated and honorable century-long mainstream of research begun by Alphonse Laveran. These last words of what has happened might be viewed as sensationalism—an unworthy departure from a narrative that has dealt with the biomedical and public health issues of disease and death of tropical peoples. But we must remind ourselves that today the administration of research may have an enormous influence on the scholarly work "at the bench." We must remind ourselves that malaria, by sheer numbers, remains the most important disease of humans—100 to 200 million case year that bring death to 1 to 2 million. We must remind ourselves that the measures now available to relieve the burden of malaria from tropical peoples are woefully imperfect and inadequate. We must remind ourselves that the AID malaria Vaccine Project is the largest, most lavishly funded program ever mounted to find a new method to deal with malaria—and that from this project there have come six indictments charging the manager, scientists, and affiliates with theft, conspiracy, criminal solicitation, and tax evasion. Nothing like this has ever happened in any other enterprise of scientific research.

In his pre-malarial professional life as a research veterinarian, Miodrag Ristic had found that when blood parasites of cattle, *Babesia,* were held in short-term culture, they would secrete a soluble proteinaceous substance into the medium. This soluble protein was antigenic and could be used as a vaccine to protect cattle from the serious and economically important disease caused by *Babesia. Babesia* is a protozoan related to the *Plasmodium* of malaria. The possibility that the secretory antigen of the malaria parasite could also serve as a vaccine was worth exploring, and AID gave Ristic money to carry out preliminary studies in experimental animal malarias. However, the trials in experimental animals were disappointing and there was no contemplation of proceeding to tests in humans.

In 1983, Ristic maintained that his line of research was still worthy of support and he submitted a proposal to AID for a further three years funding. His budget was $2.38 million. AID by this time had established an expert panel of consultants who reviewed the proposals submitted for funding and were to periodically assess the progress of the research of those projects that were funded. The advice of this panel was not only frequently ignored, but its opinions were falsified by AID's malaria vaccine project officer James Erickson when he submitted documentation to the "treasury," the Office of Procurement, to request the disbursement of money. Such was the case with Ristic. The expert panel recommended it not be funded because it was more like a preproposal. AID's malaria vaccine project director, Erickson, went to his office of Procurement with the fiction that the expert panel "had endorsed the scientific methodology and the exceptional qualifications and experience of the researchers" (GAO report).

Ristic got his millions, but there was little or no further progress toward the development of a vaccine. If the research was unrewarding, the funding was not. Charges of fiscal improprieties were leveled against Ristic. The University of Illinois's audit charged that from 1984 to 1987, Ristic transferred approximately $24,000 to his personal account from travel tickets he purchased with AID grant money. The university associate chancellor for public affairs stated that "Ristic purchased airline tickets using grant funds, then he didn't end up taking the trip. He had credits for unused tickets deposited into accounts with travel agents which were under his personal control."

The University of Illinois allowed Ristic to return the money and to take an honorable retirement of emeritus status in which he retained the rights to his office and laboratory space. This didn't satisfy the investigators of AID's Office of the Inspector General, however, who began their own audit. In 1987, those investigators began to look at the account books. They deemed that there had been sufficient fiscal improprieties to warrant criminal investigation. The evidence was turned over to the Department of Justice and the Attorney General of Illinois. A nervous University of Illinois severed all ties with Ristic in February 1990 and ordered him to vacate his laboratory and office space. Four months later, in June, he was indicted (and now awaits trial) on four counts of theft and theft by deception. The Illinois Attorney General, Neil Hartigan, characterized it as "blatant and shameless greed." But Ristic was not alone in being charged with theft of AID research funds. Some 4,000 miles to the west, in Honolulu, Wasim Siddiqui had also been indicted, accused of siphoning off $130,000 through illicit accounting tricks.

Wasim Siddiqui was able to exploit two major discoveries of other scientists' malaria research. The first breakthrough was made in 1966 by Martin Young of the National

Institutes of Health, who showed that the South American owl monkey *(Aotus trivergatus)*—a small, endearing, great-eyed creature—was susceptible to experimental infection with the human benign tertian malaria parasite, *Plasmodium vivax.* The following year, Quentin Geiman of Stanford University proved that the owl monkey could also be infected with *Plasmodium falciparum.*[73] It was now possible to conduct vaccine screening trials with the "real thing," or at least, half of the "real thing"—the parasite was "human" but the monkey was still an experimental animal. Furthermore, some dissident researchers argued that the infection in the owl monkeys was quite unlike that in humans.

The other breakthrough occurred in 1977, when William Trager and James Jensen of the Rockefeller University discovered a method of growing *Plasmodium falciparum* continuously in culture. The critical factor that had eluded everyone else all the past years was ridiculously simply—except that no one else had thought of it. Trager and Jensen realized that instead of needing more oxygen to grow, the falciparum malaria parasite, maturing and reproducing exclusively in the deep recesses of the body's vasculature, was actually in a habitat of reduced oxygen tension. They put the falciparum in a rich culture medium containing fresh human red blood cells, lowered the amount of oxygen, and raised the amount of carbon dioxide in the incubator. Now the parasites grew and reproduced indefinitely in the cul-

73. Earlier, it had become known that the higher apes—gorillas, chimpanzees, and splenectomized gibbons—were susceptible to experimental infection with human malaria parasites. These animals are, of course, too scarce and expensive to be used commonly for experimental purposes. Old World monkeys are totally insusceptible to infection with human malaria parasites. Why a monkey of South America, a continent that almost certainly had no malaria in pre-Columbian times, should be susceptible to human malarias remains a biological and evolutionary mystery.

ture vessel so long as fresh medium and red blood cells were periodically added. So far, only *Plasmodium falciparum* and a few monkey malaria parasites have been successfully cultured, but that has been enough; the most deadly of the human malarias, the primary target for vaccine development, could now be harvested in bulk from the culture flask.

Siddiqui immunized owl monkeys with a vaccine made from *Plasmodium falciparum* cultured by the method of Trager and Jensen and mixed it with Freund's adjuvant. The vaccinated owl monkeys were protected. After challenge, the infection was suppressed to low numbers of parasites in the blood (while low for a monkey, that same level of infection would cause serious malaria in humans). Furthermore, without the Freund's adjuvant, the monkeys couldn't be immunized at all and *with* the Freund's adjuvant, the monkeys developed the customary soreness and lesions. Like all the previous monkey immunization results, it was not a vaccine that could ever be used in humans.

Nevertheless, Siddiqui's AID-funded study was publicized as if human vaccination was a surety in the future. Press conferences were called. The State of Hawaii legislature voted its gratitude to him. The University of Hawaii honored him with its most prestigious award for excellence in research. This was not only for Siddiqui's discovery of "finding the first promising candidate for a malaria vaccine" but also for his being the first to culture malaria parasites and the first to infect owl monkeys with human malaria.

These misrepresentations were so outrageous that they evoked protests—including one from Dr. Leon Rosen, a former president of the American Society of Tropical Medicine and Hygiene. The award was given despite the "exaggeration of accomplishments."

The search for a malaria vaccine based on the blood (asexual)-stage parasites began to shift by the mid-1980s. Greater "purity" was demanded. Whole parasites such as those employed by Silverman and Siddiqui were no longer considered to be suitable as vaccine candidates. The "new" science was commendable as far as technology and concept were concerned, but it was still shooting in the immunological dark. It was unknown which, if any, of the myriad of parasite antigens would elicit a protective, immune response. Wonderful ways were applied to make a "purer" vaccine—isolation by binding to monoclonal antibodies, recombinant gene techniques, and, like the sporozoite vaccine, synthetic fabrication of the antigen. AID continued to be the major source of funding for malaria vaccine research, but others in Australia, the United Kingdom, the Netherlands, and Colombia took up the quest for the "pure" vaccine.

Contrary to theory, the "purer" the antigen, the weaker it seemed to be in protecting experimentally immunized owl monkeys. These "new age" malaria vaccines had an absolute requirement for an adjuvant, and the "purer" the antigen, the stronger that adjuvant had to be to boost the vaccine to give even partial protection to the owl monkey. And the stronger the adjuvant, the more toxic and less suitable for human use it tended to be.

To AID and its network of funded scientists, the vaccine was still around the corner despite the notable lack of success. Negative results were blurred. In one experiment in which the adjuvant under test caused the monkeys to lose appetite and weight, the author concluded that the adjuvant "may be safe for human use." The published paper would describe "minor discomfort" of the adjuvant-injected monkeys while the animal attendants would privately describe crippled animals.

Because the blood-stage antigens were too weak to use alone and too toxic to use in combination with adjuvant, they were never tested in humans. Nevertheless, in 1986 Siddiqui was to write that "the asexual blood stage vaccine may soon be available for clinical trials". But in 1986, twenty years after AID's entry into the vaccine program, there was no vaccine. In 1986, AID was at $63,779,000—and spending.

Despite the lack of progress, in 1984 or 1985 Siddiqui submitted a proposal to AID to extend his research for another three years, at a cost of $1.65 million. AID sent the proposal to two external expert referees. Their evaluations, cited in the GAO report, were:

Reviewers 1 said, "The proposal is mediocre, overly ambitious and the budget is overwhelming and excessive."

Reviewer 2 said, "The proposal is unrealistic in terms of time, money and availability of material. The amount of money requested is outlandish and outrageous."

The expert evaluation was again totally ignored by James Erickson. Siddiqui got his money. All of it. More "vaccines" were made. More monkeys were vaccinated and more claims made that the vaccine was almost ready for human use. More press conferences were called. Siddiqui's car carried a vanity license plate celebrating a fatal disease—MLARIA.

AID has its own independent investigative division, the office of the Inspector General (OIG). In 1988, acting on "information received," the OIG began investigating Siddiqui's and the University of Hawaii's handling of the malaria vaccine research funds. The OIG investigators' report stated that there was an apparent systematic diversion and theft of funds, as well as the submission of false claims and other documents intended to cover up the actual use of the funds that were under the control of the university's Principal Investigator (Siddiqui): ". . . there is evidence to support the allegations that the Principal Investigator apparently diverted to his and his secretary's personal use [funds] in excess of $50,000. An additional $10,000 were used to refurbish his offices at the University and apparently these construction costs were charged to the Grant as consultant payments."[74]

That was only the beginning; the final amount that Siddiqui is charged with purloining is approximately $130,000. This is certainly not an "Ivan Boesky" amount, but in science, where honesty governs all that one does, the moral equation makes it equal to a Wall Street defalcation.

On September 14, 1989, the Grand Jury in Hawaii brought in an indictment charging Siddiqui and his administration assistant, Susan Lofton, with theft in the first degree, criminal solicitation, theft in the third degree, and criminal conspiracy. Eighteen months later, after a series of delays, they still await their day in court. Lawrence Goya, Hawaii's Deputy Attorney General, charges that money was stolen by a variety of means. Some of it was siphoned off through a kickback arrangement with a Honolulu travel agency. In 1984, AID paid for an Asia Pacific Conference on Malaria that was to be convened in Honolulu in April 1985. It was a big conference and it cost a lot of money. Siddiqui was the organizer and custodian of the conference money. The AID conference check was deposited with the Research Corporation of University of Hawaii, which was charged, at Siddiqui's direction, to pay the bills. Meanwhile, Siddiqui had engaged a travel agency as agents for the conference and told the Research Corporation to give them

74. Cited in the GAO report, p. 24.

$100,000 as an advance payment for their services and for a deposit to the Pacific Beach Hotel, where the conference was to be held. In March 1985, a month before the conference, it is charged that he instructed the travel agency to begin paying him directly $1,260 a month and to pay his secretary $1,000 a month. These monthly salary "supplements" were to continue for the next two years.

When the conference was over, $18,030 remained in the Pacific Beach Hotel account. According to the investigators, $17,139 came to Siddiqui for his use. Then he sent AID a bill for $35,425, which he claimed was for his services and departmental rental related to the conference. AID paid. The indictment disputes that the money was used for that purpose.

The AID Inspector General investigators began to close in. Senator Dan Inouye of Hawaii, chairman of the Appropriations Committee, became suspicious and loosed a General Accounting office investigation on Siddiqui and the entire AID malaria vaccine program. To elude being brought down, the indictment charges that Siddiqui went to the Comptroller of the Research Corporation and persuaded him to sign backdated letters which stated that the monies taken were authorized and for legitimate expenditures. The Attorney General called that criminal solicitation.

Even during the criminal investigation of his fiscal activities, Siddiqui had an uncanny ability to raise money. Two month's before Siddiqui's arrest, the Rockefeller Foundation's health division, gave him a $75,000 research grant. Then on the day, the very day, that Siddiqui was arrested by the Honolulu Police, acting on evidence given them by the AID Inspector General's office, the AID Malaria Vaccine Research office announced that it was giving Siddiqui $1.65 million to continue the research and that Siddiqui would continue to be in charge as principal Investigator.

One of those infuriated by the award was Senator Inouye, who went on television to say that if Siddiqui handled any federal funds, he personally would see to it that the University of Hawaii would never get another cent of federal research money. AID reacted by directing that Siddiqui be disassociated from all of the projects's activities. It directed the university to find an *acceptable* replacement as Principal Investigator to take scientific and administrative charge. The university replaced Siddiqui with a man named Satoru Izutsu, who is a psychologist. In the shadow of this malariopsychologist the real workers that AID had found acceptable to carry out a $1.65 research program were a recent Ph.D. from Hong Kong and a bacteriologist with a modest publication record. In 1990, the then AID project administrator, an Army colonel, said he liked them because they were "enthusiastic."

If Siddiqui's affairs seem not really worthy of the ingenuity of a Ph.D., the affairs of his mentor at AID, James Erickson, are worthy of a long-running soap opera. Erickson had been appointed the AID malaria vaccine project manager in 1982. In that same year, AID decided that it needed an independent management to oversee the project. The management contract was awarded by non-competitive selection to the American Institute of Biological Sciences (AIBS), an essentially respectable non-profit group that manages various biological and biomedical research projects for the federal government. A woman, Dorothy Jordan, was assigned by the AIBS to be project manager, and she and Erickson became great and good friends. By 1982 Erickson and Jordan were lovers. Their affair ended in 1985, according to Erickson; in 1986, according to Jordan's first version of their love story; and in 1987, according to her second version.

There was a coincidental decline in Erickson's ardor for Ms. Jordan and for the Institute which employed her. In 1986, Erickson voiced his opinion that the AIBS (and its contract manager) was incompetent, if not downright dishonest. There began a trade of bitter recriminations which brought complaints of sexual harassment against Erickson. The house of cards Erickson-AID had erected on a foundation of misrepresentations began to come tumbling down. In April 1987, his superiors[75] reassigned him "pending allegations of mismanagement." This was added to the sexual harassment allegations. And now there was anonymously supplied information of illegal activities. In October 1987, Erickson was put on administrative leave—no work, full pay—while the Inspector General investigated the accusation.

Erickson, after a year of languishing on administrative leave, took the offensive. He claimed that he was the innocent victim of a "witch-hunt." "They're trying out there to find something on me. If I had done anything really wrong I would have hid and run." He sued the AIBS through a whistle-blower statute that allows federal employers to seek justice from erring civilian contractors. In July 1988, he legally enjoined AID to demand that it come to a decision—fire him or put him back to work. This time the court

75. Erickson's immediate superior-supervisor for the malaria vaccine project was himself! On the table of organization his immediate superior was the AID Chief of Communicable Disease. Erickson the agricultural entomologist was made Chief of Communicable Disease. "That's a wonderful vaccine research proposal," said Erickson the vaccine project manager to his boss, Erickson the Chief of Communicable Disease. "That's a good enough recommendation for us. We'll fund it," said the AID Office of Procurement.

agreed and said that AID must get off the pot by October 6. AID's response was to dock him a week's pay for bad judgment and personal misconduct. That was all it could do. Nothing had been proven. AIBS's lady, Dorothy Jordan, refused to press a formal charge of sexual harassment.

Erickson should have taken his own advice and run. The Inspector General continued to probe for evidence of a criminal offense and on November 29, 1989, Erickson was indicted by a federal grand jury. An account of the indictment in *The Washington Post* noted only the charge that he had given false information to his superiors; specifically that he had falsely told the office of Procurement that the KT&R Laboratory in St. Paul, Minnesota, had been recommended for a research grant by external reviewers.[76]

It seemed such a flimsy basis for a criminal indictment. There was a general feeling in the scientific community that AID had vindictively turned on its favorite son for making such a hash-up of the failed (but still operating and moneyed) malaria vaccine project. What Erickson had done to the project through mismanagement, amateurism, and misrepresentation was considered to be wicked but not

76. The GAO report described the KT&R transactions as follows: "In 1985, AID awarded a 3-year $736,801 grant to KT&R Laboratories, St. Paul Minnesota, to conduct the work outlined in its proposal. Unfavorable pre-award reviews of the proposal were reported as favorable, in documentation sent by the MIVR project officer [Erickson] to his superiors and in documentation sent by the office of Health to the Office of Procurement. This grant was terminated in 1987 after (1) the Office of Health evaluated subproject performance and found major deficiencies and unreasonably high expenditures given the subproject's limited accomplishments and (2) OIG audited subproject records and found KT&R's accounting system to be totally inadequate. More than $430,000 had been disbursed under this award." The OIG investigated allegations that neither action was an "arms length" transaction and that the project officer and the researchers acted improperly. The OIG turned over evidence developed during these investigations to the Department of Justice.

criminal. There seemed to be no evidence that Erickson had ever personally profited—that he had ever received a kickback from KT&R. The people who knew him wondered why he would have gone to such lengths to ensure that an outfit like KT&R would receive AID money.

Then a final indictment was brought in by the grand jury. *The Washington Post* story gave an imperfectly synoptic account of the charges leveled against Erickson. The indictment charged him with conflict of interest, conspiracy, illegally accepting gratuities, making a false claim, and submitting false income tax returns. Erickson stood accused of making illegal profits, through a tangled web of schemes, from the AID malaria vaccine research project which he managed.

To begin with, the relationship with KT&R was not an innocent one marred only by mismanagement and misrepresentation. Erickson stood accused of making money from KT&R. The story that unfolds from the indictment is that in 1985 Erickson, after approving a research proposal of $736,801, advised KT&R to enter into a contract with International Insect Research and Development (IIR&D), based in Guatemala, for technical and advisory assistance. This in itself was peculiar because KT&R's research proposal was to develop a simple immunological diagnostic test for malaria that could be carried out in a physician's office, whereas IIR&D, as its name would imply, had entomological expertise. The indictment accuses Erickson of having a financial interest in IIR&D and profiting from the arrangement with KT&R. KT&R was found to have paid $72,000 into IIR&D's Washington, D.C., bank account and Erickson made withdrawals, of unspecified amounts, from that account. In addition, KT&R paid $16,000 directly into two bank accounts held personally by Erickson.

The story of the monkeys that made men rich and contributed to the decline and fall of James Erickson forms yet another of the bizarre episodes in the AID malaria vaccine project's history. Sometime during 1983 or 1984, AID decided that a large number of candidate vaccines suitable for use in humans would shortly be ready for testing. These vaccines would, it was believed, have to be tested in the owl (*Aotus*) and the squirrel monkey (*Saimiri*, another South-Central American monkey which can be infected with human malaria parasites) for safety. This was a self-delusion; there was no effective vaccine then (or now) suitable for humans, and as far as safety tests are concerned, the FDA doesn't demand monkey trials as an absolute prerequisite to experimental use in humans.

The AIBS, still the management arm of the project, was to arrange for the purchase, and Erickson recommended that a George Diaz be given a contract to act as "headhunter" to find and acquire the monkeys. Diaz and Erickson then contacted Matthew Block, president of Worldwide Primates, based in Miami, who said he could supply two hundred owl monkeys for $475 each and four hundred squirrel monkeys for $375 each for a total of $245,000. Diaz then told AIBS that the monkeys were to be obtained from a company called Gerrick International for $336,000—$630 for an owl monkey and $525 for a squirrel monkey. Erickson, the AID project manager, approved the purchase and authorized AIBS to issue checks for payment to Gerrick International.

But who or what was Gerrick International? Gerrick International was Diaz and Erickson. The indictment states that on August 26, 1985, Erickson ordered stationery and envelopes bearing the letterhead "Gerrick International" from a printer. The order was collected September 12 and on that same day George Diaz and his wife opened a bank account in the name of Gerrick International. The AIBS check for $168,000, partial payment for the monkeys, was deposited to the account. Gerrick International paid Worldwide Primates $122,500.[77] The indictment then says that in January 1986 Diaz wrote a check on the Gerrick International account for $8,500, payable to his brother-in-law, Leonel Rosales. A week later Leonel Rosales issued a check for $8,500 payable to James M. Erickson. In March, George Diaz withdrew $11,886 from the Gerrick International account and with the money bought two cashiers checks, for $6,880 and $5,000 respectively, payable to J. Erickson. The grand jury considered that this warranted the criminal charges of conspiracy, conflict of interest, and accepting a gratuity.[78] No income tax was paid on these profits and so three counts of submitting false tax returns were added to the indictment.

Erickson never came to trial. After the long protestations of his innocence, on February 2, 1990, according to *The Washington Post*, he pleaded guilty to the charges of accepting an illegal gratuity, making a false tax return, and making false statements. He could have been put in durance vile for up to five years and fined $250,000. A lenient court sentenced him to six months, to be served in a half-way house, and fined him $20,000.

Meanwhile there was the problem of actually obtaining the monkeys—six hundred owl and squirrel monkeys are not easy to come by, even from the "home" countries whose

77. On October 4, 1985, AIBS paid Gerrick the remaining $168,000; presumably a profit was made on this also. The indictment accuses Erickson and Diaz of making a total profit of $54,250 on the resale of the monkeys.

78. George Diaz was also included in the indictment, charged with conspiracy. He is believed to have fled the country.

governments cooperate in their capture. Worldwide Primates-Gerrick International were having a difficult time filling the order for which they had been paid.

Owl and squirrel monkeys come from Colombia, Bolivia, and Peru. In Columbia, AID gave that country's National Institute of Health a $1.53 million, three-year malaria vaccine research grant. It was recognized that the Institute was not capable of doing the research and that the real purpose of the money was as a "sweetener" to get Colombian owl monkeys.[79] The monkeys were shipped. The Institute kept three sets of books, and $147,000 of the grant money turned up in the personal Swiss bank account of one of the Colombian scientists, Carlos Espinal. Espinal has been indicted for fraud by a U.S. District Court, but there is as much chance of getting Espinal to trial as AID has of getting its money back for the hundreds of surplus owl monkeys now languishing in substandard quarters in Florida.[80]

The Peruvian deal was, if anything, even more cynical than that of the Colombian monkey connection. AID gave $1.1 million to Peru, ostensibly for a program to *conserve* the owl monkey in the wild. Then it had Peru ship six hundred owl monkeys, for which an additional fee was paid on each animal.

The Bolivians were the first to react to the depredation

79. Malaria vaccine research seemed to attract bizarre connections to crime—even to the shadowy world of drug interdiction in Colombia. During one of Erickson's trips to Colombia, an important diplomat asked if AID couldn't spread the defoliant of Vietnam notoriety, Agent Orange, on the coca trees under the guise of an antimalarial spray program. Erickson, according to the account in *The Scientist* of July 10, 1990, replied that the scheme was irrational and impractical.

80. There is a more extensive account of this story in the July 29, 1988, issue of *Science*, pp. 521–523.

of their monkeys by AID. Their press published articles expressing their indignation and the Bolivian government put an embargo on the export of owl monkeys. Worldwide Primates' president, Matthew Block, attempted to smuggle a cargo of "wetback" owl monkeys. He was caught, threatened with prison, and fled in a private plane. Colombia and Peru followed Bolivia in making it illegal to export *Aotus*, but by that time fifteen hundred animals had already been sent to AID. Many of these animals were of the wrong subspecies and quite useless for malaria research. Other animals had no experimental place to go—there were no vaccines to test. Furthermore, there was a groundswell of criticism that the *Aotus–Plasmodium falciparum* system was an unreliable representation as a model of human malaria. Meanwhile, the monkeys have been in "storage" at a cost of about $2 million.

And another unaccountable AID action. By 1987 it was clear that there were no candidate vaccines safe enough or worthy enough to be tested for field trials in humans. Yet, it was in 1986 or 1987 that AID decided it required a facility in an endemic area to test the phantom vaccines. In some ways this made sense, and such a facility should have been established not long after the project began. The monkey trials could only provide an imperfect prediction of what might happen in semi-immune humans. Even the screening in American "immune virgin" volunteers would be unreliable to predict the potency a candidate vaccine might possess when applied to the village setting. However, the "village trial" in a sample community would require at least a year's preliminary study to obtain the baseline demographic and epidemiological data. Establishing a field research station was one of the few aspects of the AID malaria vaccine project that made sense, but it came at a time when the project was scientifically bankrupt and on the verge of collapse.

There were those within AID who argued against establishing a field station. Nancy Pielemeyer of AID's Policy and Planning Office stated that "Field stations to test vaccines would be established even though no clinical successes are in hand and no candidate vaccines have been tagged for development. This is not a reasonable foundation for a $23 million project."

You would think that the tropical nations which would be the chief beneficiaries of a vaccine would have offered their collaboration in providing facilities and subjects for field trials. There were no such offers. Africa and Asia said America first and foremost. Siddiqui approached an Additional Secretary (Undersecretary) for Health of India to have the human vaccine trials there. Despite the intractable problems with malaria in India, the government official rejected the blandishments. The Indians were not going to be experimental animals—especially for an American vaccine. Later, when the vaccine did not materialize and Siddiqui had fallen from grace, that (former) official congratulated himself on his perceptive wisdom.

At last, Papua New Guinea was made an offer it couldn't refuse. In a government-to-government agreement, AID was to award approximately $20 million for a five-year field project. This budget was later cut in half and the time extended to eight years. The actual research was to be carried out by the Papua New Guinea Institute of Medical Research, a first-rate establishment with a long history of applied and fundamental research on malaria in the field and laboratory.

The population designated for any eventual vaccine field

trial consists of villagers not far from Salata. There are now staff from the Institute in the field, mapping and making censuses of the villages, taking blood films, and feeling spleens. Entomologists are capturing mosquitoes, identifying their species, and examining them for the presence of sporozoites. A medical geneticist is attempting to characterize the immunogenetic constitution of these Wosera, East Sepik, people. Work goes on, awaiting the delivery of a candidate malaria vaccine that may never come. But at least some meaningful, honest investigations are being done for the Yankee dollar.

After twenty-five years, the AID malaria vaccine research project has proven a disaster, and by promising more than it could deliver to the expectant Third World, it may have done actual harm. However, to give AID its due, there was / is a true need for a vaccine, even if used only as an adjunctive measure to control endemic malaria. AID stepped in when other, more "respectable" scientific establishments ignored that need and abandoned their responsibilities.

AID failed because it was run by amateurs who would not heed the advice of professionals. AID failed because it succumbed to sleaze and corruption. AID failed because it fostered mediocre science and over-inflated the meaning of experimental results. It may also be that AID failed because the human constitution is such that no vaccine can confer a protective immunity.

However, it is possible that the villains were not the indicted, but the respectable, established, and honored scientists. These were the men and women who said not a word in public protest when their opinions were either ignored or manipulated into falsifications. These were the men and women who said in private that the AID-sponsored research was of doubtful quality. These were the men and women who disregarded their responsibilities as leaders of their profession. Their silence may have caused irreparable harm to the future of malaria research. The malarious are still with us and they still need help.

Index